swim to recovery
canine hydrotherapy healing

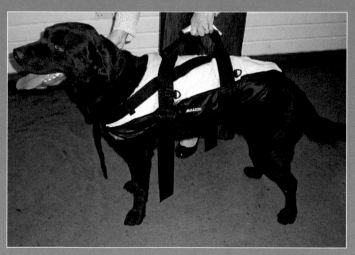

Hubble & Hattie

Emily Wong

Gentle Dog Care

The Hubble & Hattie imprint was launched in 2009 and is named in memory of two very special Westies owned by Veloce's proprietors.
Since the first book, many more have been added to the list, all with the same underlying objective: to be of real benefit to the species they cover, at the same time promoting compassion, understanding and co-operation between all animals (including human ones!)
Hubble & Hattie is the home of a range of books that cover all-things animal, produced to the same high quality of content and presentation as our motoring books, and offering the same great value for money.

MORE TITLES FROM HUBBLE & HATTIE

A Dog's Dinner (Paton-Ayre)
Animal Grief: How animals mourn (Alderton)
Cat Speak (Rauth-Widmann)
Clever dog! Life lessons from the world's most successful animal (O'Meara)
Complete Dog Massage Manual, The – Gentle Dog Care (Robertson)
Dieting with my dog (Frezon)
Dog Cookies (Schöps)
Dog-friendly Gardening (Bush)
Dog Games – stimulating play to entertain your dog and you (Blenski)
Dog Speak (Blenski)
Dogs on Wheels (Mort)
Emergency First Aid for dogs (Bucksch)
Exercising your puppy: a gentle & natural approach – Gentle Dog Care (Robertson & Pope)
Fun and Games for Cats (Seidl)
Know Your Dog – The guide to a beautiful relationship (Birmelin)
Miaow! Cats really are ncer than people (Moore)
My dog has arthritis – but lives life to the full! (Carrick)
My dog is blind – but lives life to the full! (Horsky)
My dog is deaf – but lives life to the full! (Willms)
My dog has hip dysplasia – but lives life to the full! (Haüsler)
My dog has cruciate ligament injury – but lives life to the full! (Haüsler)
Older Dog, Living with an – Gentle Dog Care (Alderton & Hall)
Partners – everyday working dogs being heros every day (Walton)
Smellorama – nose games for dogs (Theby)
Swim to recovery: canine hydrotherapy healing – Gentle Dog Care (Wong)
The Truth about Wolves and Dogs (Shelbourne)
Waggy Tails & Wheelchairs (Epp)
Walking the dog: motorway walks for drivers & dogs (Rees)
Walking the dog: walks in France for drivers & dogs (Rees)
Winston … the dog who changed my life (Klute)
You and Your Border Terrier – The Essential Guide (Alderton)
You and Your Cockapoo – The Essential Guide (Alderton)

WWW.HUBBLEANDHATTIE.COM

First published in October 2011 by Veloce Publishing Limited, Veloce House, Parkway Farm Business Park, Middle Farm Way, Poundbury, Dorchester, Dorset, DT1 3AR, England. Fax 01305 250479/e-mail info@hubbleandhattie.com/web www.hubbleandhattie.com
ISBN: 978-1-845843-41-0 UPC: 6-36847-04341-4 © Emily Wong and Veloce Publishing 2011. All rights reserved. With the exception of quoting brief passages for the purpose of review, no part of this publication may be recorded, reproduced or transmitted by any means, including photocopying, without the written permission of Veloce Publishing Ltd. Throughout this book logos, model names and designations, etc, have been used for the purposes of identification, illustration and decoration. Such names are the property of the trademark holder as this is not an official publication.
Readers with ideas for books about animals, or animal-related topics, are invited to write to the editorial director of Veloce Publishing at the above address.
British Library Cataloguing in Publication Data – A catalogue record for this book is available from the British Library. Typesetting, design and page make-up all by Veloce Publishing Ltd on Apple Mac.
Printed in India by Imprint Digital Ltd

Contents

Acknowledgements • Preface •
Introduction • Animal Magic. . . 4

1 Ethics & welfare. 10
   Animal welfare. 11
   Animal ethics. 14

2 Hydrotherapy: in the
   beginning 16
   Hydrotherapy & people 18
   Horses. 19
   Canine hydrotherapy 19
   Underwater treadmill. 21

3 Hydrotherapy: an
   introduction 23
   Pre-hydrotherapy assessment. 26
   The patient 29
   Breeds. 29
   Age. 30
   Ill health. 30
   The introduction. 31

4 The magic of water 34

5 The aches & pains of arthritis. . 39
   Osteoarthritis 40
   Rheumatoid arthritis 41
   Septic arthritis 42

Case history: Katie. 43
Case history: Sonny 45
Case history: Dotty 46

6 Obesity: a big problem 49
   Age. 51
   Reproduction & hormonal
   abnormalities 52
   Genetic predisposition. 52
   Voluntary activity level 52
   External control of food intake 52
   Diet type, amount, frequency 53
   Disease & disorders 53
   Environment & lifestyle. 53
   Weight management 54

7 Hip dysplasia 55
   Clinical signs 56
   Prevention scheme 57
   Diagnosis. 58
   Surgical management. 58
   Case history: Alfie 59
   Case history: Wally. 62
   Case history: Rory 64
   Case history: Ralph. 67

8 Elbow dysplasia. 70
   Diagnosis. 71
   Prevention 72

Ununited anconeal process 72
Fragmented coronoid process. . . .73
Osteochondrosis/osteochondritis
dissecans 73
Case history: Conker 74

9 The patella (knee cap) 77
   Case history: Archie 81

10 Cruciate injuries. 84
   Clinical signs 85
   Diagnosis and surgical
   treatment 86
   Case history: Lilly 87
   Case history: Domino 90

11 Spinal conditions 94
   Case history: Duchess 101
   Case history: Bertie. 105
   Case history: Chester 108

12 The future 111

13 Appendices. 113
   Further reading 114
   Websites .115
   Glossary .116

Index 125

Acknowledgements
Preface
Introduction
Animal Magic

ACKNOWLEDGEMENTS

A special thank you to my family and friends (mum in particular). This book is made possible by the positive publicity and achievement at Animal Magic, so much admiration goes to all of those involved – humans and animals alike.

PREFACE

This book has come about as a result of growing interest in the benefits of hydrotherapy treatment. A feature in a national newspaper about the work of Animal Magic Hydrotherapy, Fitness and Grooming Centre prompted the publisher to contact the Centre, with a view to publishing a book about its work. As an employee of the Centre, I was approached about authoring this work and tell of the joy and benefits of aquatic exercise.

This book is intended for everyone who is enthusiastic about dogs, their welfare and wellbeing, and its objective is to give hydrotherapy as a whole – and particularly as a treatment for dogs – the exposure it deserves. Little about this type of therapy has been available, and it is my aim to rectify this.

Mention of hydrotherapy for dogs usually results in raised eyebrows and blank looks: what is it, and what does it do, if anything? are the unvoiced questions. Because most of us rely on our vet to sort any medical problems with our companion animals, we may not have a good knowledge or understanding of complementary therapies, and when and where they should be used.

This book contains basic 'need to know' information on a variety of common clinical conditions and injuries, with actual case studies as examples of how hydrotherapy works, what problems it can help overcome, real, life-saving results, and an insight to the associated ethics and welfare.

Before I became a hydrotherapist and learned of its benefits, I admit I was sceptical. However, I have witnessed many cases of dogs receiving treatment, and been truly impressed with the results; after all, seeing is believing.

The objective of my book is that every reader will gain an understanding of this treatment, learn how health problems can present, the process of diagnosis and veterinary treatment, and how and why hydrotherapy works. Welfare is the main issue throughout, with emphasis on the prevalence of obesity, and medical conditions resulting from poor breeding programmes.

Emily Wong
Hydrotherapist

INTRODUCTION

The domestic dog is one of the most common household pets in western society, but an unfortunate reality is that, at some point in a dog's life, she will require veterinary treatment of some kind.

What was known as 'alternative' medicine (including physiotherapy, electrotherapy, and

Hydrotherapy is a type of aquatic exercise that's both controlled and fun.

hydrotherapy), has now been recognised as 'complementary' therapies. Conventional veterinary medicine and therapies like these work together to maximise the beneficial effects.

Double trouble; and best of friends.

Hydrotherapy is the use of water to maintain and restore health (swimming, to you and me!). It was developed from a scientific theory of hydrodynamics which was based on the properties of water, human movement, and physiology. Swim therapy, as it is also known, enables patients, be they human, horse or dog, to use a greater range of motion while carrying out weightless exercise.

Hydrotherapy can be used for a variety of conditions, including weight and pain management, rehabilitation of injuries, and pre-/post-surgical treatment. As well as being cost-effective, in some cases it may even reduce or obviate the need for surgical procedures, meaning less need to use anaesthetic (which can be risky), and a better quality of life for your friend (and your bank balance).

Hydrotherapy or aquatic exercise can help with a wide range of orthopaedic and neurological conditions, amongst others, with the objective of achieving optimal function following injury, surgery or disease.

ANIMAL MAGIC

This book is based on the work of the Animal Magic for Hydrotherapy, Fitness and Grooming Centre, which is based in my home town of Colchester in Essex, and was founded by Kerry Youngman.

Teamwork is the key to success.

KERRY YOUNGMAN

Kerry founded Animal Magic in 1992, and is proud to have become such a success and help to dog lovers in and around Colchester. Running a successful cat and dog grooming business was not enough for Kerry, whose aim in life is to help improve the quality of life of all dogs, especially by the prevention of obesity, one of the most prevalent problems affecting dogs today.

Kerry Youngman, owner and director of Animal Magic, with her beloved dogs, Grancha Sam and Molly.

Monty inspired Kerry to open her centre.

Having seen thousands of dogs during her career to date, Kerry has witnessed many cases of obesity, lameness, and arthritis, as well as hearing many heartbreaking stories. As the owner of a rescued, morbidly obese, five-year-old English Mastiff, Kerry lost the battle to restore Monty to health, despite all that she and the vets did. Over a period of time, and with the experience and awareness gained from working with dogs and her efforts with Monty, Kerry decided that hydrotherapy was the way to help as many dogs as possible.

Shortly after the Centre opened, Kerry's three-year-old English Mastiff, Grancha Sam, ruptured his cranial cruciate ligament. Using her own hydrotherapy centre to help aid his recovery, Kerry saw first-hand how amazing it really was, in the process gaining empathy and understanding of the distress that dogs and owners go through at times such as this. and what a hard struggle it is to nurse them back to health. This made her even more determined to broadcast the benefits whilst also educating dog owners about this modality.

Kerry has successfully acquired qualifications in hydrotherapy, clinical conditions, first aid, and anatomy and physiology, as well as micro-chipping, teacher training, and grooming. In February 2009, Animal Magic expanded into a hydrotherapy and fitness centre for the rehabilitation and care of dogs. Working closely with local vets, Kerry and her tightknit team have improved the lives of many dogs, and continue to do so with passion and commitment, aiming to benefit as many canines as they can!

MANDY OSBORNE

Mandy has been a valued member of Animal Magic's team since 2004, acquiring a vast amount of knowledge and experience, and gaining qualifications in cat and dog grooming.

An enthusiast and expert in canine agility, Mandy has been a member of the Tendering

Mandy with agility dog Patch.

Agility Group for four years, regularly competing and taking part in displays with her dogs, Patch and Fly, both of whom are Border Collies.

Mandy has a keen interest in hydrotherapy and assists with the dogs during their sessions. She plans to complete her hydrotherapy training in the near future.

AND A LITTLE ABOUT ME

Having grown up with small animals in my life, I aspired to not only own several pets, but to have a

Emily Wong.

career working with them, focusing on dogs as my particular species of interest.

Prior to my birth, my family owned a German Shepherd named Alex, one of my favourite breeds. Having seen photographs of him, I always longed for a dog of my own, and now am the proud owner of two, both rescue dogs. In August 2006, after much deliberation (and bribing my mother), two-year-old Milly, a Jack Russell terrier, came up for rehoming. I was overjoyed; my first dog! In September 2010, I rescued Trogg, a Lucas terrier aged three; together, they have made great companions for me, as well as each other.

I became an employee at Animal Magic after completing six months of work experience back in 2006, when I was studying for a National Diploma in Animal Management at Writtle College. After completion of my work experience, I stayed on at the Centre and gained a BSc (Hons) Degree in Animal Management. Prior to the opening of the hydrotherapy centre, I completed a Pool Water Management course with the Canine Hydrotherapy Association at Hawksmoor Hydrotherapy of Excellence. Whilst working in hydrotherapy I became aware of the lack of research about this subject, so studied the effects of canine hydrotherapy for my dissertation, focusing on hip dysplasia in particular.

As a graduate, I went on to achieve a Level three Diploma in Grooming. After shadowing and assisting Kerry with hydrotherapy for almost a year, I completed a hydrotherapy theory and practical training course with the National Association of Registered Canine Hydrotherapists at Greyfriars Rehabiliation Centre.

There are several more qualifications to acquire, and additional training courses to carry out. My objective is further study of animal welfare, canine behaviour, and physiotherapy, and I hope to complete a Masters degree in the future.

My ultimate ambition is to be involved with animal welfare and husbandry in Hong Kong, amongst other countries, where I'd like to raise awareness through education about the correct and humane treatment of dogs in all their different

Milly, the Jack Russell Terrier.

Trogg, a Lucas Terrier.

roles. I hope that, as a result, positive changes can be made to improve the future of dogs everywhere.

Indeed, there is much work to be done within the United Kingdom in terms of improving animal welfare and dispelling misconceived ideas about dogs, the latter usually the result of sensationalist media coverage. In situations that involve injury to humans, it is the dog that is blamed and demonsied, when, in many cases, the reality is that we are to blame because of ignorance and poor management.

Not taking crossbreeds into account, there are over one hundred and fifty recognised dog breeds. Whatever we deem their role and purpose in life to be, it should always be remembered that they – and all creatures – deserve to be treated with respect and understanding.

After all, they are man's best friend ...

1
Ethics and welfare

Ethics and welfare are the cornerstones of animal care. As a responsible dog owner, you will always take into consideration the best interests of your dog. Unfortunately, dogs experience illness and injury, just as we do, and it falls to you to decide on treatment and recovery options when this happens.

Hydrotherapy has many advantages and benefits in this respect, as is discussed in the next chapter, but it has some limitations, too, and can be contraindicated. When considering hydrotherapy as a treatment, a full understanding of its pros and cons are necessary to decide whether it is the right form of therapy for your dog. Moreover, hydrotherapy can only be truly effective when used in tandem with care at home.

This is where ethics and welfare come into play. Most owners will do anything for their dog to ensure he has the best life possible. However, at what point does this become for our benefit rather than our dog's? How far would you go to extend your dog's life, and is it right to do so? It is important to understand the consequences of our actions and be careful not to act selfishly out of love.

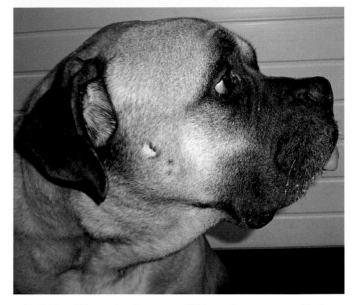

Bull Mastiffs and other mastiff breeds are more likely to experience respiratory problems due to their 'squashed-in' facial features.

Animal welfare

Humans bear the responsibility for the welfare of all animals. Animals are chosen to join a household for various reasons, such as companion, therapy dog (guide dogs, etc), or to work in some way, and it is the carer's duty to ensure that their animal's health and welfare are taken into account. To preclude ill-health, good welfare involves an animal being physically and psychologically fulfilled. The UK's Animal Welfare Act 2006 specifies the welfare needs which should, in theory, ensure that animals are protected from harm. The UK's Animal Farm Welfare Council formulated the 'Five Freedoms' commonly used as a benchmark for animal welfare. With respect to the dog, the majority

Shetland sheepdogs, Border Collies, and Greyhounds have dolichocephalic skulls (the head is disproportionately long and narrow).

Lucas Terriers are a cross between Norfolk and Sealyam Terriers, both of which have moderate skull proportions.

Crossbreeds are at less risk from developing hereditary conditions.

are denied at least one of the five freedoms, according to reports in journal papers.

THE FIVE FREEDOMS

- Freedom from hunger and thirst – by ready access to fresh water and a diet to maintain full health and vigour

- Freedom from discomfort – by providing an appropriate environment, including shelter and a comfortable resting area

- Freedom from pain, injury and disease – by prevention or rapid diagnosis and treatment

- Freedom to express normal behaviour – by providing sufficient space, proper facilities and company of the animal's own kind

- Freedom from fear and distress – by ensuring conditions and treatment which avoid mental suffering

Dogs were originally bred to perform specific tasks, such as hunting and herding. Since then, other considerations – such as lifestyle and personal preference, or competitons/showing have come into play.

An animal's welfare can be termed 'poor' when its physiological systems are disturbed to the extent that survival or reproduction is impaired. Selectve breeding means that specific anatomical features have been exaggerated in many breeds with little regard to health, welfare, disposition and their ability to function.

The domestic dog (Canis familiaris) originates from the Grey wolf (Canis lupus), which, when it is compared to some of the dogs of today, highlights that the extreme morphological changes which have resulted from selective breeding mean a reduced quality of life in conjunction with high incidences of inherited disorders. In some cases, such as with the Shar Pei, Pomeranian and Pug, certain physical features have been exaggerated to such a extent that the dog experiences pain

and suffering purely from the business of living. Respiratory disorders affect behaviour, because the animal is unable to communicate or behave naturally. The English Bulldog is a classic example of this: selective breeding has resulted in the brachycephalic skull (nearly as broad from side to side as from front to back) which makes proper respiration impossible. A large percentage of Bulldogs have difficulty breathing at rest, let alone during exercise, and their typical loud, laboured breathing can be perceived as growling, especially by other dogs. In addition, the breed's fixed facial expression means it is unable to physically communicate emotions, or play and display submissive behaviours. Bulldogs – along with chondrodystrophic breeds (long body and short legs), and giant breeds – have difficulty assuming a 'play bow' whereby a dog holds his hindquarters in the air with his front end flat on the ground, as an invitation to play.

The vast number of different dog breeds – which are increasing, especially with the advent of so-called 'designer dogs' – means the emergence of new diseases and disorders at a time when we are still trying to find remedies and solutions to existing ailments and afflictions. Some breeds are genetically predisposed to certain health issues, and not a single dog breed is free from some type of clinical condition, although many go undiagnosed. Nobody is perfect, and the same goes for dogs, but a healthy dog is a happy dog. Most health issues can be treated to a degree, but do we do so in order to prolong the life of our companion so that we don't have to do without him, or as a genuine response to a welfare issue?

ANIMAL ETHICS

The imperative of modern animal ethics is that animals should be protected for their own good, which places considerable demands on us humans. Empathy with animals is based on what we perceive as their human similarities, for example exhibiting a particular behaviour in response to specific expressions or responses to human emotional states. Anthropomorphism – attributing human characteristics to animals – has been used as a method of establishing animal emotion, although over-interpretation can lead to ethical and welfare issues. For instance, begging behaviour is perceived as starvation, which can result in unnecessary feeding that can cause obesity, digestive problems or even poisoning. It is vital that we differentiate between our desire to 'look after' animals, and true protection of them.

Interaction between humans and animals is the subject of much debate and opinion, the main protagonists being welfarists and animal rightists. Both camps believe that animals are sentient (have the ability to feel and perceive), and rely upon scientific research that shows how animals cope with varying experiences when they are ecologically and behaviourally deprived. Animal welfarists support humane animal use based on the idea of utilitarianism. Utilitarians believe that something is right only if the value of it outweighs any inherent suffering or harm. For instance, animal experimentation that resulted in finding a cure for cancer – even if very many animals suffered and died in the process – would be considered justifiable.

Animal rightists believe that all uses of animals (for food, experimentation, transport) is wrong, and that they should be protected from exploitation: regardless of how humane the action is, it is unacceptable.

Due to increased understanding of the welfare problems associated wth breeding, many crossbreeds – predominantly Poodle crossbreeds – are being bred to help reduce the incidence of genetic problems, as well as for better temperament and aesthetic reasons. The downside to this is that inadequate background research is leading to dogs developing a combination of both the breeds' problems, which defeats the object entirely. For instance, a Cockapoo (a Cocker Spaniel crossed with a Poodle) can inherit the ear complaints from which Spaniels suffer, and the Poodle's susceptibility to patellar luxation (dislocation of the kneecap). Careful selection of

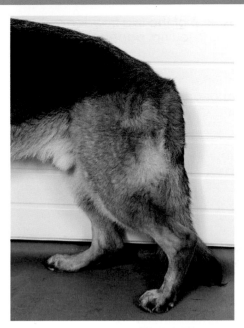

German Shepherd dogs are specifically bred to have low-set hind quarters.

West Highland Terriers commonly suffer with skin conditions.

sires and dams is one of the best solutions to such a big problem.

2
Hydrotherapy: in the beginning

Complementary medicine is comprised of therapies and modalities such as massage, ultrasound, thermotherapy, physiotherapy, and water treatments, and aims to help in acute and chronic pain management, as well as promote general health, body conditioning, and rehabilitaiton of injuries and physical disorders.

Physical therapy principles and techniques are based on a solid foundation of scientific and clinical research and adherence to evidence-based medicine principles, although there is little scientific evidence to support their effects.

Physical therapists primarily relied on physicians' guidance until the 1940s, but due to development in the profession, therapists' own clinical intuitions and experiences and/or inherited traditions were used to guide their treatment programmes.

In 1996, in its annual report of 1996, the Chartered Society of Physiotherapy (CSP) defined physiotherapy (in relation to humans) as "a healthcare profession which emphasises the use of physical approaches in the promotion, maintenance and restoration of an individual's physical, psychological and social wellbeing,

Animal Magic's hydrotherapy pool measures 2.44m high x 1.2m wide x 1.2m deep (8ft x 4ft x 4ft).

encompassing variations in health status." The CSP is an organisation which supports those who deliver physiotherapy care, and is the founder of the professional body known as the World Confederation for Physical Therapy (WCPT) which, in 2001, added that "… physiotherapy is a process that aims to facilitate individuals with impairments and activity limitations to reach their optimal physical and/or social functional level through partnership with family and the community." As physiotherapy has developed, greater understanding and skill have led to advances in specialist areas such as sports therapy and hydrotherapy.

Animal physiotherapy in the United Kingdom dates back to the early twentieth century, with specific legislation citing the application of "Physiotherapy in veterinary medicine first appearing over forty years ago." (Veterinary Surgeons Exemptions Order, 1962). Hydrotherapy, in particular, has become very popular with canine and human physical rehabilitation, as well as the equine world.

Bathing in water (balneotherapy, or spa therapy) has frequently been used in traditional medicine as a cure for diseases, and bathing in spring water as a cure for osteoarthritis, ankylosing spondylitis (a spinal condition), and to help elderly patients suffering with mobility restriction is an historic method used since Roman times. (The terms balneotherapy or spa therapy are often used only to describe passive bathing in thermal or mineral waters, and are distinguished from hydrotherapy as the beneficial effects are provided by the nature of the water, whereas hydrotherapy involves exercise in water.)

Since the beginning of the 21st century, both hydrotherapy and balneotherapy have been accepted as appropriate modalities for all forms of treatment with water, although 'spa therapy' is used as an alternative to balneotherapy. The terms water therapy, aquatic exercise, aquatic therapy, and hydrotherapy are used throughout this book, in reference to swimming exercise in water.

As already mentioned, the therapeutic use of water in all of its forms dates back to the beginning of civilisation, with treatment occurring in water that was sufficiently deep to allow for total body immersion; so in pools (man-made), natural resources (lakes, streams), and underwater treadmills. Depending on the treatment programme (passive bathing or exercise therapy), this means that exercise can be done without the need to without weight-bear.

Aquatic therapy aims to –

- enhance the range of joint motion/flexibility

- strengthen muscle

- relieve muscle spasm

- maintain and/or improve mobility

- reduce weight-bearing stresses on painful joints

- assist gait and posture re-education/proprioception (spatial awareness and position sense)

- improve respiratory function and condition of the cardiac muscle and cardiovascular system

- alleviate pain

- promote wellbeing

- assist in weight management

- promote general fitness

- provide an alternative/extra exercise method

HYDROTHERAPY AND PEOPLE

Hydrotherapy gained popularity in the mid-nineteenth century, and recent years have seen an increasing number of trials evaluating the effects of hydrotherapy when used on people, although many are still based on balneotherapy.

In a study of hip osteoarthritis patients, results showed that hydrotherapy positively improved gait in total hip arthroplasty (replacement) patients who underwent a hydrotherapy programme, post-surgery. There have been many studies based on arthritic patients, possibly because of the buoyancy effect of water, which diminishes the effects of gravity.

Hydrotherapy is also used by hospital patients with severe symptomatic disease or multiple joint conditions who find land-based therapeutic exercise too painful. In addition to physical benefits, hydrotherapy has prompted emotional states of enjoyment, relaxation, confidence, and a sense of accomplishment.

HORSES

During the nineteenth century, the use of hydrotherapy was developed to include horses. Common conditions treated include tendonitis, laminitis, lameness, injuries from competition work, and body conditioning. Traditionally, horses have been exercised in naturally cool and salty sea water, and cold water (2-4°C) is used with Epsom salts in underwater treadmills and specialised swimming facilities. Hydrotherapy is carried out in low temperatures to reduce the metabolic response of cells so that the body can operate with less oxygen, which, in turn helps prevent hypoxic injury (an inadequate amount of oxygen, causing injury), reduce the incidence of fluid build-up on joints, and act as an analgesic and anti-inflammatory.

Swimming is an unnatural activity for most horses as it is non-weight-bearing and an unusual way for equines to move. When swimming, a horse's head is often elevated, and the lumbar region of the spine held in extension (back is stretched unnaturally). As horses aren't natural swimmers, they tend to flail, which can easily result in injury;

Swimming is a natural activity for dogs, although breeding has led to conformational differences in some which makes it more difficult.

because of this, the underwater treadmill is the preferred type of treatment.

CANINE HYDROTHERAPY

Therapeutic exercise is possibly the most significant form of treatment in small animal rehabilitation in cases of neurological and musculoskeletal conditions. It can also be used for leisure and fitness purposes as swimming can require physical stamina which benefits the whole body.

Early hydrotherapy for dogs took place in lakes, rivers, and the sea. Dogs exert less heat during exercise, and are much more efficient at regulating body temperature in comparison to horses. For this reason, canine hydrotherapy is carried out in heated water (26-30°C) and swimming in natural waters is not recommended

The underwater treadmill allows all of the patient to be seen during treatment.
(Courtesy Sue Hawkins, Hawksmoor Hydrotherapy Centre)

as, even in summer months when the water has been heated by the sun, it generally remains too cold. Cold temperatures cause the body's blood vessels to constrict and send the blood around the vital organs and away from peripheral surfaces and limbs. As a result, muscles and limbs become cold and injury is likely. Other disadvantages of swimming outside a hydrotherapy centre is that the dog cannot be monitored accurately (cloudy water), and potentially harmful items such as

broken glass and fishing tackle may be in the area. The main concern, of course, is the risk of drowning due to lack of control, no buoyancy aid, or exhaustion.

Professional hydrotherapy for dogs is administered in a pool or on an underwater treadmill, both of which provide a safe environment for planned and structured active exercise of appropriate intensity and frequency for each patient.

Underwater treadmill

Walking in water has become a common form of rehabilitation hydrotherapy for humans and animals, and is thought to be particularly beneficial for injuries of the limbs, and conditions affecting the spine where proprioceptive retraining (spatial awareness) is required. The underwater treadmill is ideal for use with post-operative dogs that require a slow re-introduction to weight-bearing. Dogs standing on (dry) land bear all of their bodyweight (100 per cent); in an underwater treadmill a dog can experience a reduction in weight-bearing via adjustment of water levels.

Several different models of underwater treadmill exist, and all have the same basic features. There are usually two doors (one at each end to create a 'walk way'), and a ramp (for entry and exit). The underwater treadmill has a rubber, non-slip belt for the patient to walk on, with side panels that allow the hydrotherapist to safely stand within the treadmill. This is important because it enables him or her to provide assistance, support, and encouragement to the patient throughout the session. Clear glass around the treadmill allows outside viewing of the patient from all angles. Two hydrotherapists should work on one patient to ensure complete control of the treadmill environment, and monitor the patient by watching and physically assessing her.

The underwater treadmill is operated remotely, and temperature, water level and speed can be regulated as required. Varying water depth means that different muscles and areas of the body can be 'worked.' Unlike hydrotherapy, the treadmill allows water pressure and resistance to be increased or decreased, and thus tailored to a specific treatment programme.

Water level is always set above the bony landmark that is being treated. For example, for a

Most underwater treadmills have two doors so that the patient doesn't have to turn around once inside. (Courtesy Sue Hawkins, Hawksmoor Hydrotherapy Centre)

cruciate injury, the water level would be set above the stifle (knee), because if it was set at the point of the knee, the dog would experience too much pressure on and around this area, which would effectively do more damage than good. With water at tarsus (hock) level, the limbs would bear 91% of the dog's weight; at stifle level, it would be 85 per cent of body weight; and at hip level, the figure drops to 38 per cent. Varying water depth allows a gradual progression of weight-bearing and support of the body.

The underwater treadmill and hydrotherapy pool can be used in conjunction to create a highly beneficial programme. Although the underwater treadmill can be more strictly controlled, the pool is the main benefit to hydrotherapy overall.

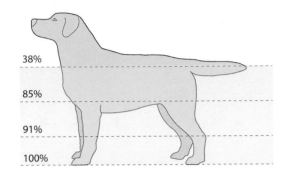

Water levels can be adjusted to the individual patient's needs. Percentages show the amount of weight-bearing relative to water level. (Courtesy Dave Russell)

3
Hydrotherapy: an introduction

The Canine Hydrotherapy Association (CHA) is a regulatory body that represent the majority of hydrotherapy centres in the United Kingdom. The CHA has set out standards and Codes of Practice which all members must strictly follow to ensure quality of care within their own centre.

One feature of becoming a member of the CHA is that at least one employee per centre undergoes thorough training and obtains specific credentials to ensure treatment excellence. In 2009 Animal Magic became a full member of the CHA and continues to achieve its benchmark.

The National Association for Registered Canine Hydrotherapists (NARCH) is an organisation whose committee works hard to become a recognised body in the United Kingdom. It maintains a list of

Regular testing of the pool water ensures the highest water quality ...

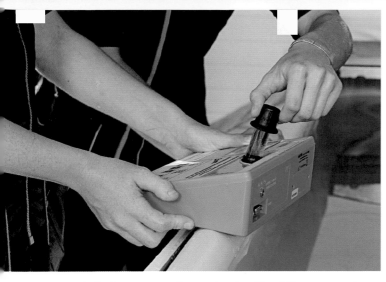

treatment by a veterinary surgeon prior to being introduced to the water. It is vital that vets health check and assess the patient to ensure she can undergo aquatic exercise. Even if the only reason for hydrotherapy sessions is that the dog's owner wants her dog to swim simply for the fun ot it, and the fitness benefits, the dog could possibly have an undiagnosed, life-threatening health issue

... and photo comparator water testing kits are most commonly used for this.

registered canine hydrotherapists, and Animal Magic is in the process of becoming a member of NARCH.

Several courses exist for hydrotherapist training, some of which are aggregate, such as First Aid, pool water management, and hydrotherapy itself (practical and theory). The subject of hydrotherapy, as with a lot of therapies, is a constant learning curve, and as more research is carried out on its benefits and effects, new and rare clinical conditions are being discovered. As a hydrotherapist, you never know what conditions you are going to be dealing with, but the minimum requirement is a general, all-round understanding of the most common disorders, diseases, and injuries. All dogs must be treated as individuals, and some methods will only work with particular characters and conditions. Experience is essential, and in many cases it is the dog who teaches the hydrotherapist how to refine handling and techniques. And in this job, it's true that you do learn something new every day!

All patients must be referred for hydrotherapy

Water therapy is only carried out on a veterinary referral basis.

that hydrotherapy could bring to light. A classic example of this would be a heart or skin condition. Generally speaking, dogs that have not been examined by a vet within six months of requesting hydrotherapy must do so and be given the all-clear. However, it is in the patient's best interest to also be checked within 24 hours of hydrotherapy treatment, and it is obviously important that hydrotherapists and vets liaise with one another to discuss any areas of concern, progress and medical history.

On arrival at the relevant centre, hydrotherapy patients are re-assessed to verify that they are in good enough form to swim. Even veterinary referrals do not guarantee that patients undergo aquatic therapy if, for example, a dog is suffering from acute pyrexia (fever), vomiting and diarrhoea, eye, ear or skin conditions, or showing a general deterioration in condition. To ensure that their dog is ready for hydrotherapy, owners are given guidelines to help with the smooth-running of the session. Dogs should not have eaten a meal or drunk a lot of water for a minimum of two hours before a session, and should not eat for two hours after hydrotherapy either. Owners must ensure that their dog has relieved herself prior to the start of treatment.

Pre-hydrotherapy assessment

This assessment is carried out on all patients in order to understand and get to know what is 'normal' for the individual. It also helps to detect any changes in condition, improvement of the ailment, and general progress (weight, muscle development).

The assessment is carried out in a methodical order from head to tail. Ten main points are checked, in addition to noting any injury or condition –

- eyes
- ears
- mucous membranes
- skin

WEIGHT

PRE-HYDROTHERAPY HEALTH ASSESSMENT

DATE	WEIGHT IN KILOGRAMS	COMMENTS/ ADVICE GIVEN, CONDITION SCORE (1-5)

BODY MEASUREMENTS

BODY PART	MEASUREMENT	COMMENTS
LEFT HIND LIMB		
RIGHT HIND LIMB		
WAIST		
TORSO		

TEMPERATURE:

PULSE:

RESPIRATION/ BREATHING RATE:

GAIT (WALK AND TROT, STRAIGHT LINE, TURNING LEFT AND RIGHT, STANDING)

COMMENTS

OTHER COMMENTS (Areas of concern, findings on palpation)

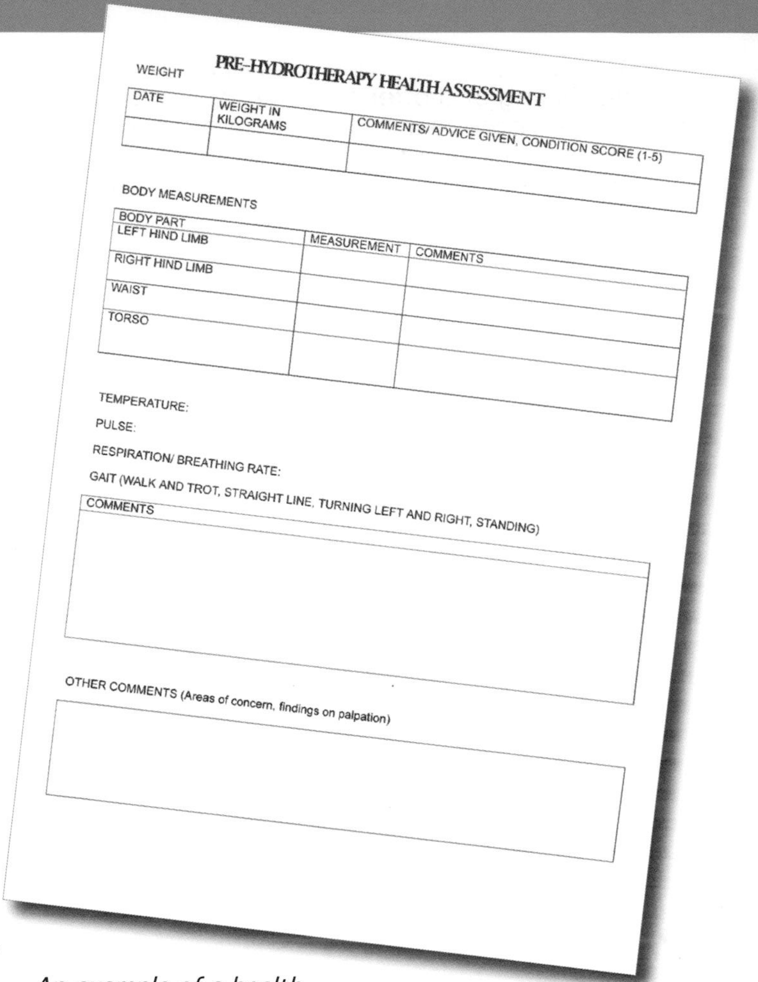

An example of a health assessment form, which should be completed prior to hydrotherapy treatment.

- weight
- body measurements
- gait
- pace
- temperature
- pulse
- respiration

All information is recorded in as much detail as possible. During and after hydrotherapy sessions, the patient is checked regularly in comparison with these notes to highlight any cause for concern, in which case, the session is stopped immediately.

● Eyes
A healthy dog has alert, bright, reactive eyes that may secrete some clear discharge. The patient should not exhibit abnormal discharge or redness, be squinting, have an eye closed, sores, inflammation or swelling.

● Ears
Ears come in an array of shapes and sizes, and some dogs are more prone than others to ear infections; Spaniel breeds, for example. The ear should be felt and checked for heat, swelling and any soreness. The patient should be assessed to determine whether she would benefit from a 'swim hat' to prevent water entering her ears.

● Mucous membranes and capillary refill time (CRT)
These are lubricating membranes that line all body openings such as the mouth, nostrils and eyelids, and are usually a light pink colour. The mucous membranes of the mouth are most commonly checked for hydrotherapy. Dark pigmentation and individual colour variations do occur, so it is important to accurately describe what is seen.

During treatment, it's vital that mucous membranes, breathing, and behaviour are regularly observed and checked.

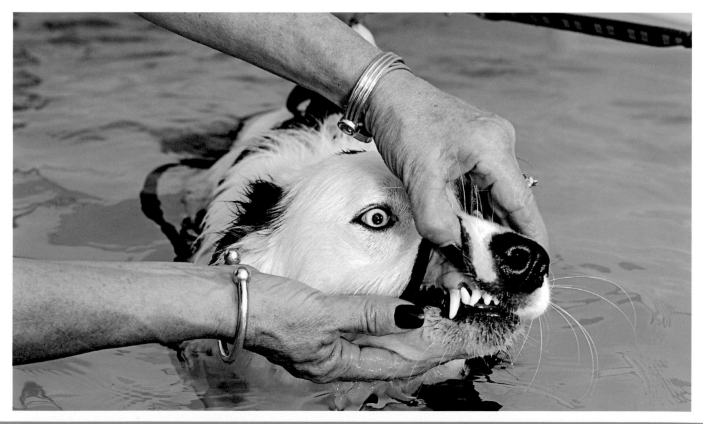

A deep red indicates over-oxygenation of the blood, which is likely to happen after exercise. Over-exercising can lead to heat stroke, so regular checks of the membranes during the session will prevent this.

The capillary refill time is a test that is used as a quick assessment of shock or circulatory failure. Pressure is applied to the upper gum, above the canine tooth, which causes the gum to blanch, and the time it takes for the gum to return to its usual colour is measured in seconds. A normal capillary refill time is 1-3 seconds; any longer than this indicates that the patient is suffering one of the above-mentioned states.

● Skin
The skin should be examined over the entire body to check for sores, inflammation, lesions, (open) wounds, irritation, parasites, etc. Chlorine can irritate skin and cause changes to it.

● Body measurement and weight
To ascertain the progress made with aquatic exercise (ie muscle development and weight loss), measurements are taken of the girth of the dog's quadriceps, torso, and waist. In combination with regular weight checks, overall body condition can be recorded.

● Gait
Gait, or patterns of movement, can be divided into symmetric and asymmetric locomotion. Asymmetric locomotion includes the gallop, a high speed, almost running movement, which, due to its physical demands and general inappropriateness for hydrotherapy, is not used for assessment.

Symmetric gaits include the walk, trot, and pace, which all involve the limbs on one side of the body mirroring the motion of the limbs on the opposite side. (An asymmetric gait like the gallop is when the limb movements of one side do not mirror those of the other.) When walking, a dog usually has three feet touching the ground at a time, or possibly two. The pace is even and steady because weight is distributed evenly. A patient with

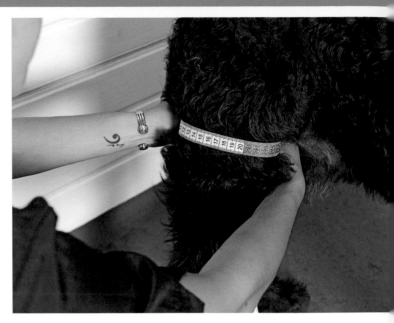

As part of the health assessment, measurements are taken so that changes in muscle mass and body fat can be noted and compared in reassessments.

proprioceptive (spatial awareness) and balance issues will appear unsteady and gangly.

During assessment, patients are required to walk in a straight line as well as in circles (clockwise and anti-clockwise), unless it may cause them discomfort; for instance, in post-spinal surgery cases. At a walk, any lameness may be subtle and hard to spot as the sound legs provide support and stabilise movement. Lameness is more likely to be evident at a trot, when just two feet are in contact with the ground. The diagonal pairs of limbs (the left foreleg and right hind leg are a diagonal pair) move almost simultaneously. This type of gait is faster than a walk, and requires quicker transition of weight distribution. Often, in the case of injury, dogs will hold up a limb and completely or partially non-weight-bear during a trot to protect the affected limb.

● Pace

This is when lateral pairs of legs move almost simultaneously (the right front leg and right hind leg are a lateral pair), and this occurs when a dog moves the fore limb and hind limb on one side while bearing weight on the other. This can be commonly seen in dogs with short bodies and long legs (such as the Whippet), and is an attempt by the dog to avoid kicking or tripping over itself.

With a change of pace, the distribution of body weight alters, and in lame dogs this transfer may be exaggerated or uneven. A really painful condition or injury may mean that the dog does not weight-bear at all on one of his legs, and will hold it off the ground when still.

● Temperature, pulse and respiration

It's likely that a patient's stress levels will be raised on arrival at the centre, so this should be taken into account when assessing her. The normal temperature for a dog is 38.3°C (100.9°F) to 38.7°C (101.7°F), with a pulse rate of 70-100 beats per minute and respiration rate of 15-30 breaths per minute.

The most common and easiest pulse to find is the femoral artery, located on the inside of the femur. A fast pulse could indicate that the tissues are not getting enough oxygen, which the heart compensates for by beating faster to meet the body's requirements. Even though an increase in pulse rate is expected during and just after exercise, over-exertion and stress can cause an unwanted increase.

Respiration, or breathing rate, can be described as slow, normal or fast. Breathing rate can be checked by watching the rise and fall of the chest to time the movements: one rise and fall equates to one breath. Flared nostrils, laboured breathing, excessive panting, a head and neck that are extended in an effort to obtain more air, breathing through the mouth, and abnormal breathing sounds all indicate a lack of oxygen. If this should occur, the patient should immediately be removed from the pool.

Some respiratory sounds are normal, depending on breed and any known condition, such as laryngeal paralysis (an inability to dilate the space between the two vocal chords at the entrance to the larynx), in which case, even closer monitoring is then necessary to recognise the difference between what is normal and abnormal.

THE PATIENT

All dogs must be considered individually with regard to treatment, taking into account their breed and temperament, and also likes and dislikes. For instance, some dogs are food-orientated, whilst others prefer toys/specific toys. A dog may have a nervous disposition or fear of strangers, maybe people with glasses, for example. The owner plays a key role in helping staff get to know their dog as they can provide much information about his personality and normal behaviour. Taking in and digesting information such as this from owners, and combining it with their own initiative and experience as a hydrotherapist is vital in recognising and dealing with a variety of dogs. Owners spend much more time with their dog than does a hydrotherapist, and will pick up on changes, however small and seemingly unimportant. An example could be a dog who seems reluctant to go out for a walk because it is cold and raining, when, in fact, he is actually in pain and doesn't want to move to minimise the pain he is feeling. Another example is a dog who, instead of retrieving his ball as many times as you are willing to throw it, does this a couple of times only and then stops. Is this because he is bored with the game ... or maybe there's a health problem, or joint or muscle discomfort?

BREEDS

All dogs are able to swim; it is a natural behaviour going back to their ancestors of the Grey Wolf. But, due to evolution, some breeds are now less equipped – and less inclined – to swim than others. Take the Labrador and Pomeranian as examples. The Labrador Retriever was bred to retrieve game from lakes and rivers, and does so admirably and with little effort. The Pomeranian has the same anatomical features as a Labrador, but was bred to fulfil the role of lapdog, and so his reduced size, double-thickness coat, and shortened nose make swimming a much more challenging prospect.

AGE

Dogs of all ages can undergo hydrotherapy, depending on their general health and any pre-existing conditions. If swimming for general health and fitness maintenance purposes, they should generally be at least six months of age, although this is breed- and dog-dependent. Larger breeds mature at a much slower rate than small dogs, so exercise should be managed carefully to prevent over-exertion, damage to immature joints, etc, and excess muscle growth. Increasing muscle mass is one benefit of hydrotherapy, but for young dogs whose bones have not yet properly developed, the muscle mass can become too heavy for the bone to support.

ILL HEALTH

Aquatic exercise is not possible if a dog –

• is incontinent

• has diarrhoea and vomiting

• has open wounds

• has skin infections (acute moist dermatitis or hot spots, skin disease)

• has a contagious disease such as parvovirus, or a zoonotic disease (can be communicated between dogs and humans) such as leptospirosis, ringworm, or sarcoptic mange

Hydrotherapy, generally speaking, can be carried out once a wound has healed, and in some post-operative cases, it is started at two weeks after

Flotation devices like this foam head ring help support the head, and prevent ears getting wet.

surgery. Following joint surgery, however, it is not advisable for four to six weeks because of the greater stresses placed on the joints.

Some conditions make it too risky to swim, in particular those involving the cervical region (neck) of the spine (see chapter 11), frequent epilepsy, and undiagnosed shoulder problems. If water therapy is used incorrectly, it can do more harm than good; establishing the source of an injury or condition is vital though in some cases does remain unknown, in which case the dangers posed by this far outweigh any possible likely benefit.

Dogs with cardiac and respiratory dysfunctions should be swum with caution, and depending on veterinary diagnosis; again, it may endanger

Buoyancy or non-buoyancy jackets are used to swim each patient.

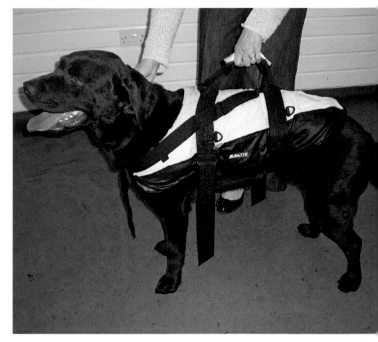

the dog. Similarly, a dog suffering from von Willebrand's disease, anaemia, or any condition that compromises blood supply, won't be able to produce the oxygen necessary for cardiovascular exercise such as this.

Hydrotherapists should always treat every patient with caution, and some dogs may need to be monitored even more closely, because of potential issues with obesity, brachycephalic breed types, laryngeal paralysis, hyper/hypoadrenocorticism disease, diabetes, and heart murmur. Liaison with each patient's vet is vital for comparing observations and medical notes.

THE INTRODUCTION

Each patient starts the session wearing a life jacket depending on the individual and his requirements, and treatment begins with a short spell in the water. Dogs that have difficulty standing ordinarily can usually do so in the water, as partial weightlessness reduces pressure on painful limbs, making hydrotherapy ideal for dogs with lameness or post-operative pain. There is a vast selection of buoyancy aids for dogs from life jackets to swim hats, and all equipment must be CHA (Canine Hydrotherapy Association) approved.

During assessment, the correct type of buoyancy aid is chosen and fitted. Some dogs are used to wearing coats, etc, whilst others have never had that experience. All dogs react differently to new stimuli, and many are unfazed by a life jacket, although it's not unusual for for a dog to happily allow a buoyancy aid to be fitted, only to then refuse to move on land with it on. It's obviously very important that patients become accustomed to wearing a buoyancy aid such as a life jacket.

Entry and exit to and from the pool must be carefully considered in order to minimise stress, prevent injury, and ensure continuity of the

session. How this is done is mainly decided by the general condition of each patient: whether there is lameness or immobility, with particular regard to post-operative care. In addition, temperament and disposition (nervous, stubborn, or anxious) are taken into account.

Animal Magic offers three entry and exit options: hoist, ramp, and manual lifting. The hoist is generally not required a great deal, but is available if necessary (for dogs with severe spinal injury, which rules out the possibility of manual lifting), as well as for heavy dogs.

Manual lifting is the most common form of entry and exit, especially with small patients. Two hydrotherapists are in attendance during sessions at Animal Magic (although only one is required to swim a dog) to ensure that careful lifting techniques are used; particularly important with large and/or nervous dogs.

To swim a patient, a hydrotherapist needs to be able to observe, time, and encourage (by massage, moving through the water, vocal encouragement, playing with toys). From experience, all dogs prefer that their owner is nearby, although sometimes this can ellicit unwanted behaviour, and prove a distraction. Owners are briefed about their contribution to the session, and asked to remain calm and relaxed, as dogs sense stress and anxiety, especially from their owner. This negative energy can be reflected in the patient's performance and affect his experience of the treatment.

Dependent on the individual, patients usually need a hydrotherapist in the pool with them to support, direct, and encourage them. Being in the pool allows the hydrotherapist to examine the patient whilst in motion, and encourage or discourage incorrect movement. Many dogs are swum by being held in the middle or to the side of the pool with the hydrotherapist outside and handling them from the top. This allows the same amount of physical examination as when the hydrotherapist is in the pool, and is particularly good for panicky dogs as it means they can retain their personal space, and are not travelling around the pool, which sometimes causes stress.

Using the ramp to enter and leave the pool is the best option for active and capable dogs, as they can walk into and walk out of the pool with minimum handling, and also for larger dogs, and those who are naturally confident and inquisitive. Nervous patients often find this method stressful as they're unsure of what lies ahead. As they acclimatise to the pool environment, many patients develop more independence and accept the ramp; toy-motivated dogs are keen to use the ramp if it means they can get to a toy. The ramp not only acts as a pathway to and from the pool, but also a platform where patients can rest without the need to leave the pool and return a while later.

Aquatic exercise is demanding on the cardiovascular system. Dogs with low fitness levels, who use fast swimming strokes, or have brachycephalic head shapes require several resting periods during a session. Conversely, some patients have such high stamina that, given the choice, they would rather continue swimming than rest. Respiration rates, mucous membranes, and capillary refill times are regularly monitored and checked to ensure that the patient is not over-exerting himself, and rested as and when the hydrotherapist feels it is necessary.

Precisely how long a patient is in the pool depends on his response to the exercise, prior activity away from the centre, age, health status, behaviour, lifestyle, and regularity of sessions. Aquatic therapy is said to require four times as much effort as land-based exercise, so a five minute swim is equivalent to a twenty minute brisk walk. If a dog is highly active at home, hydrotherapy could over-exert him, so it's up to the owner and hydrotherapist to balance and adjust exercise regimes accordingly.

Dogs attending the centre two to three times a week have a different swim duration allowance to patients who attend once a week, in consideration of their treatment programme. Patients who are over-exercised by hydrotherapy or any activity can suffer from Delayed Onset Muscle Soreness (DOMS), whereby the body has a delayed reaction to physical activity and adopts this method to cope with it. Affected dogs may experience stiffness and swelling around joints, muscle tension, lameness, and exhaustion, resulting in the need to sleep. Once-weekly patients can swim for longer as they are able to recover before the next session, but a patient who has four, 20 minute sessions a week will suffer fatigue of both body and mind if not allowed adequate resting time in-between sessions.

A hydrotherapist must take into account, together with all other considerations, at what rate swimming is done and using what technique. As previously mentioned, all dogs can swim; it is a natural ability, although many breeds are not 'programmed' or physically adept at this type of exercise. Almost all dogs will swim with encouragement and patience.

Unfortunately, the pace or rate at which a dog swims cannot be controlled. Over-excited and hyper dogs often swim quickly, which causes the body to work exceptionally hard against the water. However, big, slow strokes similarly require extra strength because of the resistance encountered.

The hydrotherapy pool at Animal Magic is fitted with anti-swim jets at its top end. These are adjustable to enable an increase in current, making it more challenging for a dog to swim against. Anti-swim jets are only used with those patients who have exceptional fitness levels, and require an intense hydrotherapy programme, for example, working dogs. Anti-swim jets can be used for strength training, stamina development, and body conditioning. Some dogs with severe muscle atrophy (wasting) on one side/limb can be swum at an angle so that the affected body part is swimming against the jets, and thus working much harder than the rest of the body, which is not directly in the jets but still having to resist the current. However, gradual progression is required with this otherwise the patient would be unable to cope.

Once dogs have had their hydrotherapy, it's essential they are showered in untreated (normal) water to rinse away the chlorine (or bromine). Drying patients with professional free-standing, dog grooming dryers is important to prevent skin complaints caused by moist, warm skin, as well as to ensure that the patient is comfortably warm and dry.

Each time a patient visits the centre, detailed information about his activity at home, and his progress or deterioration since his last visit is taken into account, and he is weighed and examined before entering the pool. Details are noted on a

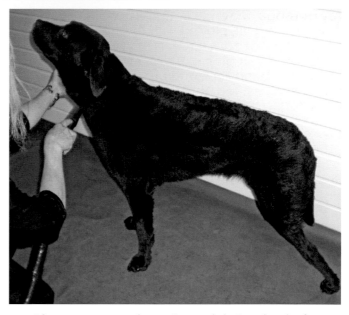

After treatment, showering and drying the dog's coat helps prevent skin complaints and irritation.

swim record to remind the hydrotherapist of what has gone before, and allow her to make informed decisions about the progress of the case.

A lot goes into treating a dog in this way; it's not just a case of placing a dog in water and allowing him to swim about. A controlled and well-thought-out method for hydrotherapy treatment under guidance is what results in positive outcomes for the patients.

4
The magic of water

Hydrotherapy practice has been scientifically developed based on the theory of hydrodynamics, the branch of science that deals with the dynamics of fluids, especially incompressible fluids, in motion.

Hydrotherapy has been used to treat disease and injury (in humans) in many different cultures, including the Egyptians, Persians, Greeks, Hindus, Chinese, and Native Americans. Knowledge and understanding gained from immersing humans in water has allowed aquatic therapy to be used as a method for facilitating movement and function in animals.

The temperature of a canine hydrotherapy pool should range between 26 and 30°C. Warmth applied to the body allows an increase in blood flow, as well as promoting relaxation and providing pain relief, all of which contribute to an overall improvement in function and/or performance, and a more effective and comfortable form of exercise. As more blood flows through and around the body, it supplies more oxygen. Of course, this means that if the water is too cold, the cells constrict, and the blood rushes to the vital organs and circulates these to help maintain internal body temperature. As a result, the extremities do not receive an

adequate blood supply and become cold, making them prone to injury, stiffness, and worse.

Care should always be taken when swimming a patient that is stressed or working hard, as hyperthermia (abnormally high temperature) is a potential risk. This can be a problem in summer due to hot car journeys to and from the centre, for instance. A basic knowledge of the principles and properties of water are essential in order to help maximise treatment and rehabilitation programmes.

The key properties included in this chapter are –

- Specific gravity and density

- Buoyancy

- Viscosity

- Hydrostatic pressure

- Surface tension

DENSITY

This is a measure of how much mass is contained in a given unit volume (density = mass/volume). Densities of various substances are defined by specific gravity. The relative density and specific gravity of an object determine whether it will float or sink. If the proportion of an object is more than 1, the object will tend to sink, whereas if the ratio is less than 1, the object will tend to float. Therefore, a patient with lean muscle that is not moving will have a tendency to sink faster than an obese dog. Improperly selected and placed flotation devices can unbalance the patient, causing him to tilt or flip over. To maintain balance, flotation devices can be used to reduce the effort of the lean dog to stay afloat. However, a non-buoyancy jacket may be required for an obese patient as he is more likely to float.

BUOYANCY

The buoyancy of water diminishes the effects of gravity. The body is supported in the water, which

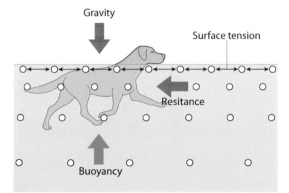

The water surface is more viscous, making it more challenging for small dogs to swim through.
(Courtesy Dave Russell)

allows for physical exercise with less stress on joints, bones and muscles, and increases independence of movement. Buoyancy enables weightless exercise which means an animal in water is effectively less heavy than on land. Patients that are weak, unstable, or have joint pain will be able to move more freely in the water with considerably less pain due to the lack of weight-bearing on joints. Water depth, movement in water, and the use of flotation devices all contribute to buoyancy.

HYDROSTATIC PRESSURE

The deeper in water that a body is immersed, the greater the pressure placed on it. When the body is still or at rest, pressure is applied equally all over. Hydrostatic (water) pressure prevents the occurrence of blood and oedema (swelling) in the lower regions of the limbs during exercise, because it increases blood circulation. Water pressure may also help minimise pain during exercise by stimulating skin receptors which reduce the pain response received by sensory receptors.

Care must be taken with patients with respiratory conditions as water pressure affects the lungs by making breathing harder. Simply submerging a patient in water can cause the breathing to become laboured (ie panting); a physiological reaction to the pressure of the water on the ribcage. In this instance, it may be several sessions before there can be any actual swimming to allow the patient to become acclimatised to the pool environment.

VISCOSITY

Viscosity is resistance of a fluid to flow. Viscosity is greater in water than in air, so water can provide resistance that may strengthen muscles, improve cardiovascular fitness, and enhance sensory awareness. Viscosity decreases as water temperature rises, thus, hydrotherapy pools set at the correct temperature will relax and help reduce soreness, whilst allowing weaker muscles and parts of the body to move more easily.

A variety of movements in water may increase or help reduce resistance. Streamline flow is the

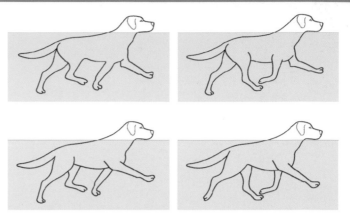

Large dogs will be more affected by hydrostatic pressure as their extremities are deeper in the water than those of small dogs (Courtesy Dave Russell).

steady, continuous movement of water, during which minimal friction occurs between layers of water as the layers separate to move around an object. Conversely, turbulent flow is irregular movement of layers of water; for example, swimming in circles makes water direction change, creating waves. Irregular movement causes elevated friction between surface tension molecules and the water, increasing resistance. Resistance in aquatic exercise can be increased by enhancing the surface area of the patient or body part moving in the water (ie a non-buoyancy aid promotes an increased body surface). Resistance to movement is also slightly greater on the surface of the water because of the effect of surface tension, which is the tendency of water molecules to adhere to each other, forming a 'skin' on the water surface. Consequently, small dogs are required to swim through much greater viscosity as their bodies are nearer the water surface. Small dogs will tend to swim really quickly in comparison to a large dog, which has to move more slowly due

Anti-swim jets are used to increase programe intensity.

to greater pressure deep within the water. A dog with a poor swimming technique will tend to thrash in the water, using more energy and requiring much more effort.

With a better understanding of how water works, an individual programme can be put together for each patient to maximum effect.

Continuous assessment of a patient's progress with regard to cardiovascular fitness, weight loss, muscle development, recovery, and technique will enable accurate adjustment and modification of buoyancy aids, treatment intensity, and swim duration as and when required.

5
The aches and pains of arthritis

Arthritis is inflammation of a joint, and usually several joints are affected. Simply put, a joint can be defined as the connection point between bones. At the end of each bone are articular or relating surfaces, as well as a layer of cartilage, called hyaline. There are three types of joint in the body, although only one type – the synovial joint – is discussed here. Synovial joints are associated with movement (ball and socket of the hip and hinge of the elbow). The stability of all synovial joints is enhanced by ligaments (the attachment between bones), surrounding muscles (skeletal muscles are connected to bone), tendons (which join muscle to bone), and neatly fitting bone surfaces. Arthritis is a type of degenerative joint disease which comes in three forms: osteoarthritis; rheumatoid arthritis; and septic arthritis.

Clinical signs of arthritis include –

- Lameness (especially following exercise)

- Inflammation, heat and swelling around a joint

- Pain on palpation or during exercise

- Reluctancy to exercise

- Depression

- Stiffness

- Reluctance to jump

- Muscle atrophy (loss of muscle)

- Crepitus (a crunching sensation and the sound of joint material rubbing together)

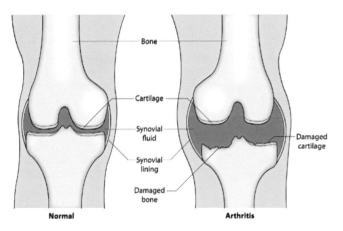

Osteoarthritis at some level is inevitable in an ageing dog. Excess weight makes it much more likely, however. (Courtesy Dave Russell)

Arthritis can be diagnosed mainly by palpation (physically feeling and manipulating the area) to assess the degree of joint changes. 'Crunching' or 'clicking of joints can usually be felt and heard when an arthritic joint is assessed. Radiographs can also be taken to observe the progress of joint degeneration. How a dog moves can also indicate arthritis. There is no relationship between the pain that an animal appears to feel and the severity of radiographic changes within a joint, meaning that arthritis could be present, but not at such a stage that it is affecting the dog's lifestyle and everyday movements.

OSTEOARTHRITIS
Osteoarthritis is the most frequently seen form

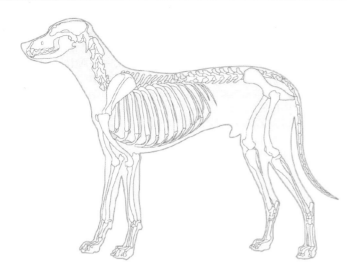

Knowledge of the canine skeleton will help ensure a thorough understanding of where conditions can develop. (Courtesy Dave Russell)

of arthritis characterised by progressive loss of articular cartilage (surface cover and protection for joints), and changes to joint and bone surfaces. Osteoarthritis is a developmental and cyclical condition characterised by joint pain, inflammation of the synovial membrane (a layer that protects the joints), and even the loss of limb function. In other words, osteoarthritis develops over time and varies in discomfort. This can show itself as lameness, which may appear to diminish and then re-occur. The condition usually gets progressively worse, although, in some dogs, osteoarthritis stabilises the joint and disease progression essentially stops whilst in others, the joint is not adequately stablisied and the ailment worsens with age.

The condition is commonly seen in conjunction with obesity, so weight control is a very important part of its treatment. Osteoarthritic dogs are more prone to weight gain, as they are less active due to the accompanying discomfort and pain, so already overweight dogs are likely to put on even more weight. Careful attention to diet is vital, along with therapeutic exercise, like hydrotherapy, to enable non-weight-bearing activity. Early treatment is necessary to prevent irreversible joint damage.

All breeds of dog can suffer from osteoarthritis. Common in geriatric dogs, young animals may also be affected due to developmental abnormalities and excessive weight. Large and giant breeds are much more likely to develop the condition due to their size (more strain is placed on the body to support the greater weight). Osteoorthritis is more likely to develop due to excessive exercise, repeated injury to the same body part, and previous history and surgery relating to joints and bone structure. Lack of adequate exercise can also precipitate osteoarthritis as muscle atrophy (wasting) occurs.

The pain of arthritis is related to synovitis: inflammation of a synovial membrane. Synovial fluid allows joints to move freely without friction, pain, or damage to the bone. Exercise therapy aims to improve muscle strength and physical function, minimising swelling and pain as a result.

To help improve the quality of life of an affected dog, anti-inflammatory medications are often administered for symptom relief. Analgesic medication helps to improve comfort and reduce lameness. Lameness occurs due to joint instability, loose ligament structures, loss of shock absorption during weight-bearing, and atrophy (wasting) of muscles.

Certain conditions that cause lameness and other conformational problems can be treated surgically. Surgical treatment focuses on correcting joint disease to prevent further joint degeneration. For example, a triple pelvic osteotomy, carried out in the early stages of hip dysplasia, helps slow the process of osteoarthritis in the hip joint.

RHEUMATOID ARTHRITIS

Rheumatoid arthritis is an immune-mediated disease (which results from abnormal activity of the body's immune system) is very uncommon in dogs. In the case of rheumatoid arthirtis, it is not known what triggers this extreme response of the immune system. A normal immune system is like a defence organisation, which goes into action when the body identifies a foreign protein by producing a response called an antigen. The body reacts to the antigen by producing antibodies which bind the antigen and destroy it, so that no further harm can occur.

In rheumatoid arthritis, the body mistakes some of its own protein for foreign protein, and produces antibodies (collectively known as 'rheumatoid factors'). These antibodies effectively attack the joint by depositing protein there, which causes inflammation. As the body tries to rid itself of the supposedly 'foreign' protein, it causes more damage to the joint. This process becomes a viscous circle, which results in the cartilage being worn away, exposing the bone and causing pain.

An affected dog will usually become lame, which may range from mild to crippling, and shift from one leg to another. Immunosuppressant and anti-inflammatory medication is prescribed to help tackle the condition, although it cannot be cured.

SEPTIC ARTHRITIS

Septic arthritis, or bacterial arthritis, is the result of an infection in the fluid and tissues of the joint cavity, and normally affects a single joint only, causing heat and pain. Pyrexia or fever, lethargy and anorexia (caused by the accompanying depression) may also be evident. Septic arthritis can lead to joint problems, rheumatoid arthritis, infection of the bloodstream, and a weakened immune system. To combat the condition, the synovial fluid is often removed. Similarly to rheumatoid arthritis, this condition is not common. Careful weight and dietary management can help delay and minimise all forms of arthritis. Regularly and gently feeling your dog's limbs will help determine whether his joints are normal. Also, visual comparison of limbs can reveal slight swelling, and touch will help identify temperature differences.

Arthritis cannot be cured or completely prevented, but the condition can be maintained. Muscle strength and general overall fitness will go a long way toward a better and more comfortable quality of life.

Maureen and Katie used to enjoy regular agility training. Unfortunately, arthritis prevents Katie from continuing this activity ... (see right).

Case history

Name: . . . . Katie
Breed: . . . . Shetland Sheepdog
Age: 7
Sex: . . . . Female
Weight (at start): 8.15kg (1.28st)
Condition: . . . . .Osteoarthritis
Owned by: . . . . .Maureen Rodwell
Surgical procedures: . . .None
Medication: . . . . .Tramadol, Previcox

INTRODUCTION TO HYDROTHERAPY

Maureen is the proud owner of two Shetland Sheepdogs, Katie and Jo, and all are keen agility and obedience enthusiasts as members of the Sudbury Dog Training Club and Gipping and District Training Club. At the end of May 2010, Maureen noticed that Katie was limping on her right foreleg. Palpation of the limb caused Katie to react strongly; she was obviously in pain. It's possible she injured herself during her agility work. Carpal joints act as shock absorbers during weight-bearing, and are prone to injury due to their lack of muscular support. Activities like agility can cause repetitive strain injury (RSI) and sprains to joint supportive tissues.

Maureen could see that Katie was keen to walk, but was experiencing much discomfort. Maureen particularly noticed that Katie seemed more in pain when she was walking on rough terrain, like shingles, and bark chippings. On 13 July, Katie was examined by the vet, and diagnosed with possible arthritis of her forelimbs. Katie's condition did not improve with administration of Previcox, and she was referred to a specialist vet for further examination.

Shetland Sheepdogs are prone to ligament degeneration, often due to rheumatoid arthritis, so this had to be diagnosed or ruled out. Radiographs were taken of Katie's forelimb, in particular her carpals and metacarpals: she was diagnosed with severe osteoarthritis. The vet advised that Katie should lose weight to reduce pressure on her joints, with exercise restricted to three, 20 minute walks on the lead a day. Katie was prescribed Tramadol as well as additional Previcox to relieve pain and reduce inflammation. In addition, hydrotherapy was advised to allow Katie to fully extend her carpals and metacarpals. Maureen was confident that hydrotherapy would benefit Katie, as she knew that the condition would only worsen with time.

AIM OF PROGRAMME

To act as conservative management for the arthritis as well as build muscle to support the right forelimb, and the body as a whole as the condition worsened. Due to the pain Katie was experiencing when walking, muscle atrophy was occurring because of restricted exercise. Muscle strengthening was important in stabilising her condition, and Katie's programme intended to help relieve the discomfort from land-based exercise; hydrotherapy was advised as the best form of exercise to combat all of these elements.

PROGRAMME DURATION

From August 21 Katie was to attend twice-weekly sessions, reducing to once a week on November 9.

TASKS TO BE COMPLETED AT THE CENTRE

To free-swim in a buoyancy jacket for up to 15 minutes per session.

TASKS TO BE COMPLETED AT HOME

No running, jumping up, down or over objects; no agility work. Maureen was advised to reduce Katie's food intake.

PROGRESS REPORT

On September 1, Katie's programme was amended

continued over

... although she still attends obedience classes, where modified commands are used to reduce physical strain.

to allow her to free-swim with particular emphasis on getting her to rotate clockwise and anti-clockwise. Previously, she was swum with the assistance of a hydrotherapist over the side of the pool to get her to extend her carpals and metacarpals correctly. Katie had progressed greatly and so a new challenge was required. She was reluctant to turn clockwise on the first few sessions of her new programme. however, and favoured her left forelimb in the pool, indicating that her right forelimb was more severely affected than originally thought, and that she was weaker on this side.

On October 7, Katie was re-examined by the vet at the specialist practice, who was extremely pleased with her progress as her stance and gait were noticeably improved. One of the main reasons for this was that non-weight bearing exercise allowed Katie to carry out an extensive range of motion, improving flexibility and muscle development. One of Maureen's goals was to get Katie's weight down to 7.7kg (1.21st),

which she more than realised when Katie's weight fell to 7.65kg (1.2st).

Katie is now able to walk for a long period of time without limping or showing signs of discomfort. Although she has boundless energy, Maureen decided not to continue agility with Katie in order to prevent further problems, athough they intend to return to obedience training in the near future. As a result of her good progress, the vet reduced Katie's medication by half a dose a day, and her hydrotherapy sessions to one a week. However, as a result of these changes, Maureen noticed Katie was not coping as well as expected, and she has returned to twice-weekly sessions, although is sticking with the lower dosage of painkiller.

In the future, it is likely that Katie will be free from discomfort for a longer period of time, and able to swim once a week.

In the pool, Katie can fully extend her forelimbs.

Even though she has a thick coat, Katie has little body fat, and requires a buoyancy jacket.

Double-coated dogs like Katie tend to get only the top coat wet.

Case history

Name:. . . . . .Sonny
Breed:. . . . . .Golden Retriever
Age: 11
Sex: . . . . .Male
Weight (at start):. 49kg (7.7st)
Condition: . . . . .Arthritis and weight
                       management
Owned by:. . . . . .Jackie Wakefield
Surgical procedures:. . . .None
Medication: . . . . .None

INTRODUCTION TO HYDROTHERAPY

Back in 2009, Jackie noticed that Sonny was becoming stiff in his gait, and showed some weakness in his back legs as if in discomfort, almost struggling to go from the down position to standing, and vice versa. Jackie concluded these were signs of old age and a stage that dogs naturally go through. Sonny was groomed at Animal Magic, and, on his next visit, staff noticed he was reluctant to remain still, and insisted on sitting down; behaviour Sonny had not shown before. Sonny was

Hydrotherapy was also advised to help manage his weight and burn off some of his seemingly boundless energy.

After veterinary examination, Sonny began hydrotherapy immediately. He suffers from an unknown ongoing skin condition which is being treated to minimise irritation. Everyone had to keep close watch of Sonny's skin after his hydrotherapy sessions in case the chlorine in the water affected it in a negative way. The condition is non-contagious, so swimming other dogs was not an issue.

Jackie is a great believer in natural and alternative therapies, and hoped that hydrotherapy would prove effective. There was no surgical option for Sonny, so hydrotherapy was worth a go if it meant there was even the slightest chance he would benefit from it.

Sonny used to lift his forelimbs out of the water, and over-extend his carpus (wrist).

As his forelimbs re-entered the water, the resultant splashing made his sessions more intense due to this resistance.

particularly sensitive to handling and lifting of his hind legs. On discussion with Jackie, hydrotherapy was recommended to help Sonny redevelop strength and regain his wellbeing; he was clearly uncomfortable.

continued over

AIM OF PROGRAMME

To improve Sonny's mobility and alleviate any pain he was experiencing from the arthritis. The intention of using hydrotherapy as a cardiovascular exercise was to build fitness levels and reduce body fat. No target weight was set, although the average weight of a Golden Retriever of Sonny's size is around 40kg (6.29st).

PROGRAMME DURATION

Sonny began hydrotherapy on June 10, 2010, attending once a week for a 5 minute sessions.

TASKS TO BE COMPLETED AT THE CENTRE

Swim in a buoyancy jacket with the ropes attached to the rear 'D' ring. Sonny lacks strength in his hind region as discomfort from arthritis has caused him to shift his body weight toward his shoulders and chest as a compensatory mechanism. To create a balance in the water, his lower back needed to be supported to allow for even body exertion.

TASKS TO BE COMPLETED AT HOME

Jackie and her mother share responsibility of Sonny, so both were advised to reduce and cut out human food and excess treats. They were also asked to keep up a regular exercise regime, being careful not to allow Sonny to overdo any physical activity.

PROGRESS REPORT

Although hydrotherapy did not adversely affect Sonny's skin condition, unfortunately, during the period 28 September to 12 October, the condition worsened, and hydrotherapy was postponed until referral from his vet was resubmitted. With the unidentified skin complaint given the all-clear, Sonny returned on October 13, attending once a week for 10 minute sessions.

Sonny has always been a very 'splashy' swimmer, due to hyperextension in his carpals (equivalent to the wrist). To help prevent this, a hydrotherapist touches the top of Sonny's front feet before over-extension occurs. As a result, he has progressed steadily, and continues with his once-weekly, 10 minute sessions.

Since his programme began, Sonny has lost 3kg (4.47lb), as evidenced by his muscle tone and ideal canine hour-glass figure. He has gained muscle mass in his hindlegs, but still needs to lose more weight. Sonny will continue swimming to maintain his fitness levels, and help manage his arthritis.

Even though, in the dog world, Sonny is considered geriatric, his weight-loss and improved fitness level mean he is more likely to live a longer and healthier life.

Case history

Name:	Dotty (aka Dotty Dream Girl)
Breed:	Staffordshire Bull Terrier
Age:	5
Sex:	Female
Weight (at start):	22.40kg (3.52st)
Condition:	Medial arthritis of the right knee and hock
Surgical procedures:	None
Medication:	Metacam; used to relieve pain and reduce inflammation
Owned by:	Mark and Jayne Hilsden

INTRODUCTION TO HYDROTHERAPY

Dotty became a member of the Hilsden family in 2005; at 6 weeks of age she was a typical boisterous and playful Staffy. In February 2010, Mark and Jayne took Dotty to the vet to get a mass removed from her left hindleg which was suspected of being cancerous. The resultant several stitches meant that Dotty was ordered to rest with minimal activity until the wound healed. Of course, being Dotty, she could not remain still despite the best efforts of

Mark and Jayne with Dotty, whose pedigree name is Dotty Dream Girl.

her dedicated owner, and her stitches burst three days after leaving the veterinary practice. Once re-stitched, the wound healed well, however, and after six weeks Dotty was allowed out for 10 minute, lead-restricted walks.

In May, Mark and Jayne noticed Dotty was limping on her right hindleg. An x-ray was Inconclusive, but the vet suspected medial arthritis of the stifle and hock. Dotty was again ordered to rest, and given Metacam, an anti-inflammatory medication.

Due to lack of exercise and muscle stimulation, Dotty lost muscle mass and definition in her leg; muscles are important to support and cushion joints, but Mark and Jayne could not help Dotty build these as it caused her discomfort to walk (hence the lameness), and she was too weak, in any case.

In addition to this, her weight was of slight concern as it was expected to increase due to reduced energy exertion in conjunction with normal food intake (extra body weight on arthritic joints causes more pressure, and heightened discomfort).

Mark and Jayne focused on maintaining Dotty's

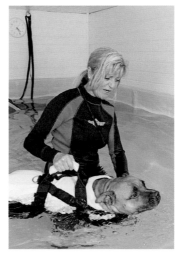

Dotty needs a therapist to control her pace as she swims very quickly.

weight, and it has been consistent throughout her treatment programme as a result.

After 8 weeks of Dotty being on Metacam, Mark and Jayne did not see any noticeable improvement in her condition, and were concerned about the deterioration in Dotty's usually upbeat temperament.

Dotty is obsessed with playing with tennis balls, and is a highly active dog. Understandably, she is not allowed to play 'fetch' in case of further injury, which seemed to frustrate and subdue her, which Mark and Jayne found upsetting.

With Dotty's mobility limited, Mark and Jayne turned to hydrotherapy in the hope that it would improve Dotty's wellbeing and return her to her old, cheeky self. There was the option to increase Dotty's medication, but after consideration, it was felt that natural healing was a far better path. As hydrotherapy is a non-weight bearing exercise, Mark and Jayne hoped that it would benefit Dotty's leg in a way that land-based exercise could not.

AIM OF PROGRAMME

To help support the arthritic leg by rebuilding muscle mass, as well as increasing mobility and range of motion.

Due to Dotty's limited exercise, her tolerance for cardiovascular exercise was greatly reduced. Hydrotherapy aimed to increase her fitness level and manage her weight.

PROGRAMME DURATION

Begun on July 20, 2010 with sessions three times a week, reducing to twice-weekly from October 25,

continued over

and once-weekly from 4 November 2010 to maintain Dotty's fitness and muscle mass. Reducing Dotty's hydrotherapy sessions per week enabled Mark and Jayne to see how well she copes with increased land-based exercise, and establish a balance between both.

TASKS TO BE COMPLETED AT THE CENTRE

Free-swim in buoyancy jacket with hydrotherapist in the pool. Swim one lap of the pool, walk up the ramp to rest; repeat as per hydrotherapist's advice, based on health and behavioural assessment.

TASKS TO BE COMPLETED AT HOME

No jumping up or down, no/limited ball games, restricted walks on the lead, reduced food intake to match energy output.

PROGRESS REPORT

When Dotty was first introduced to the pool, she could manage only two laps comfortably. She had low cardiovascular fitness, and experienced respiratory difficulties due to initial anxiety, as well as breed-related problems (her brachycephalic (short) head

Because of concern about her breathing (due to her head conformation), Dotty requires several rest periods during her sessions.

shape). Dotty swam extremely quickly and became short of breath within 30 seconds, and so was given many rest periods.

In subsequent sessions, her stamina gradually improved, which was having a positive effect at home with Dotty able to play for longer periods of time without needing to rest.

Approximately two weeks into hydrotherapy, Mark and Jayne decided to stop Dotty's medication as they felt she was not benefiting from it, and that hydrotherapy replaced the need for it.

Dotty learned to relax and swim at a slower pace, although it is in her nature to do everything at speed. At present, Dotty can confidently swim 12 laps of the hydrotherapy pool. Muscle strength and definition in her arthritic leg (and the rest of her body) has given Dotty a much more shapely figure; she is able to play ball games and walk off the lead without becoming lame.

Dotty's weight has remained reasonably consistent; however, she has lost body fat and gained muscle mass. Everyone is ecstatic at the progress Dotty has made, and she is once again playful and boisterous. She thoroughly enjoys her sessions at Animal Magic, which can be seen in her excitement when she bursts through the doors.

Dotty began swimming just once a week from November 4. The vets are pleased with Dotty's progress, and happy that she has the physical support required to move comfortably, with close observation and management of walks. Dotty's hydrotherapy sessions will increase to two a week if she deteriorates in any way. As she gets older, the arthritis will slowly progress, and is likely to occur in other joints also. Dotty will require hydrotherapy for the rest of her life for preservation of her fitness and muscle development, and to keep any discomfort to a minimum.

6
Obesity: a big problem ...

Obesity is one of the most common medical disorders in dogs, and the catalyst for many other conditions. The increase in cases of obesity is continuing; slowly but surely, we are killing the canine population and causing unnecessary suffering.

In comparison to dogs fed unrestricted diets, correct diet and exercise have been shown to reduce the prevalence and severity, and delay the onset of hip dysplasia, elbow dysplasia, diabetes, thyroid problems, and osteoarthritis, as well as many other conditions.

Obesity can be a major contributory factor in –

- Hyperinsulinema (over-production of insulin)

- Glucose intolerance (caused by hyperinsulinema)

- Diabetes (arising from glucose intolerance and hyperinsulinema)

- Cardiac disease

- Respiratory disorders

- Joint and locomotory problems (lameness, stiffness)

- Osteoarthritis

- Non-allergic skin problems (from hormone imbalance)

- Hepaticlipidosis (fatty liver disease)

- Heat and exercise intolerance (increased occurrence of heat stroke)

- Decreased immune response (higher risk of infection and disease, and longer recovery time)

- Musculoskeletal disease (in relation to the muscles and skeleton, such as hip dysplasia)

- Shortened lifespan

- Increased surgical and anaesthetic risk

- Reduced welfare and sense of well-being

A dog has no option but to depend on his owner to provide the right food in the right proportion. Starving a dog, or not feeding a dog correctly, is quite rightly considered neglect (and one of the Five Freedoms previously mentioned), but overfeeding should also be considered neglect. Too much food will have a similar outcome; problems during the dog's life and potentially a shortened lifespan.

Some dog breeds are prone to weight-gain (just as some people are), but none is immune. Clinical obesity is confirmed when a dog's body weight is 10-25 per cent above the ideal weight. Dogs within a breed type come in an array of sizes and weights; weight charts and breed description can therefore sometimes be misleading. The weight of a dog can be assessed by condition scoring, categorised into five groups: underweight; thin; ideal; heavy; and obese. Condition scoring is the equivalent to human body mass index measuring, and a general method of determining a dog's body condition relative to its age, breed, weight-to-height ratio, and muscle-to-fat proportion.

UNDERWEIGHT
- Ribs are easily visible and prominent with no fat coverage. There's a lack of muscle mass, and the waist can be clearly seen; bony hips and spine

THIN
- Ribs are clearly visible. There's a definite waist, and a minimal layer of fat covering the body can be felt. Spine and hips protrude slightly

IDEAL
- A layer of fat covers the body. Ribs can be seen and easily felt, along with the spine. An 'hourglass' figure is evident when viewed from above

Body condition score. The ideal would be a dog with an hourglass figure ... (Courtesy Dave Russell)

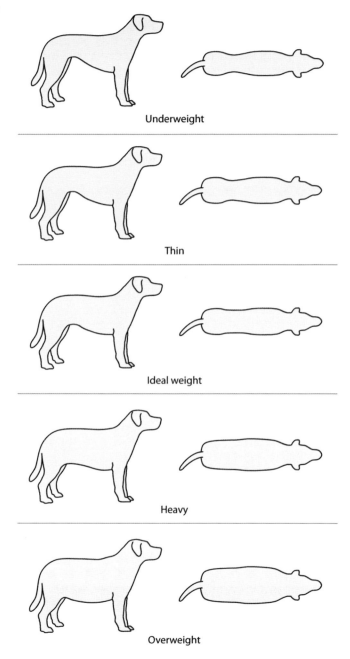

Underweight

Thin

Ideal weight

Heavy

Overweight

HEAVY

- The ribs can only be felt when pressure is applied due to a slight excess of fat. The abdomen hangs low, and a waist can hardly be defined. The hips and spine are well covered and cannot be seen

OBESE

- Ribs cannot be seen or felt. There's a high level of fat all over the body, especially around the abdomen, which is rounded. No body definition or waist when viewed from above

Obesity has been identified as the most widespread canine nutritional disorder. Correct weight and exercise management are vital to preclude a shortened lifespan. Reducing body weight can lessen the risk of related diseases, and enhance a dog's quality of life. By preventing obesity, joint stress also decreses.

Too much body fat results when there is an imbalance between energy intake and energy expenditure. However, many underlying causes also involved in the prevalence of obesity.

Factors that contribute to obesity:

AGE

Young dogs are continuously active and playful, whereas an older dog is mostly only active when necessary.

Young dogs require more nutrients to assist growth and development. The adult and geriatric dog does not need food to grow, but to maintain health and provide enough energy for usual daily activities. The older dog will exercise considerably less than the young to adolescent dog and thus burn fewer calories. A common mistake is for older dogs to be given the same amount of food throughout their lives, even though they are gradually exercising less. Similarly, puppies are

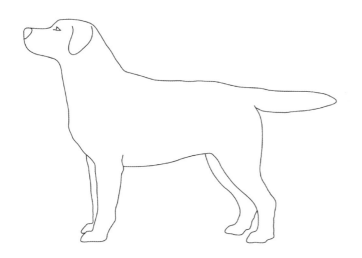

... who is lean and not carrying excess fat.
(Courtesy Dave Russell)

often overfed under the misconception that this will facilitate maximal growth, whilst growth and conformational abnormalities are more often the result.

Of course, it is easier to gain weight than it is to lose it, especially in the case of an older dog. As energy expenditure reduces with age, excess weight is much more difficult to burn off as it is necessary to exercise twice as hard and more often, meaning increased stress and pressure for the entire body.

REPRODUCTION, AND HORMONAL ABNORMALITIES

Due to a combination of physiological and environmental factors, a neutered or spayed dog is much more likely to gain weight if his diet is not adjusted accordingly.

The metabolic rate is the speed at which the body is able to utilise food for energy. Neutering and spaying reduces a dog's metabolic rate so that he requires fewer calories. Obesity has been linked to changes in hormone production as oestrogen and oestrus decrease appetite; a spayed bitch may experience an increase in appetite but her slower metabolic rate will invariably mean a weight gain. Neutering and spaying dogs does not cause obesity, but how they are cared for afterward can determine whether or not obesity will be a problem.

GENETIC PREDISPOSITION

There are links between breed type and the incidence of obesity. Most commonly predisposed dogs include Spaniel types and Labradors; of course, all dogs can be affected by obesity, although breeds such as the greyhound and whippet are genetically programmed to store little body fat, and are therefore at very little risk of becoming overweight.

VOLUNTARY ACTIVITY LEVEL (INVESTIGATIVE BEHAVIOUR, PLAYFULNESS)

Most dogs are kept as companions as opposed to working animals. Generally speaking, working dogs are much more frequently active than companion dogs, and thus get more exercise. As a companion animal, a dog may not be stimulated to be voluntarily active: searching, following scent, playing with toys and interacting with other animals. More interactive exercise is likely in households with more than one dog rather than than a dog that is on its own.

EXTERNAL CONTROL OF FOOD INTAKE (HUMAN INTERVENTION)

Unwanted behavioural habits and problems are common in relation to feeding and obesity. Human processed food should be eliminated from a dog's diet, and treats specially formulated for weight control for dogs given, as well as fruit and vegetables like carrots, apple and banana (not grapes, raisins and onions as these are poisonous). Some dogs have allergies to certain food types; ask your vet for advice.

Discourage begging behaviour by ignoring your

dog if he does this, and keeping him out of the room during human mealtimes.

Extra food and treats are usually given to show affection but, to be frank, this form of love is lethal. The least we should do for our canine companions is feed them a good, healthy diet.

Diet type, amount, frequency

Dogs are omnivores and require a mixture of meat, fruit, and vegetables to maintain optimum health. Nutrients (protein, carbohydrate, fat, minerals, vitamins, water) are necessary for the regulation of growth, repair and energy.

Dogs are naturally competitive and voracious feeders, with the tendency to gorge feed. Consequently, they cannot regulate food intake well, and need to have this done for them, although, unfortunately, it is not unknown for a dog to be fed what he 'likes' instead of what he 'needs.'

The domestic dog's ancestor, the Grey wolf, a predatory animal, in the wild, does not eat every day depending on available resources. Due to this factor, when food is available, the pack eats as much as it can as quickly as it can to preclude competition for the food, as the next meal may not be for quite some time.

As owners, we must take responsibility for our dog's diet. Food intake is controlled by complex systems within the brain that involve both internal and external cues such as smell, temperature, taste, appearance, and texture. Food that is not so highly palatable but is energy dense contains less fat and more fibre (fibre provides that 'full' feeling).

Given a choice, a dog will more than likely opt for a small portion of flavoursome, high nutrient food over a large portion of something more filling but less appealing, even though the energy dense food will be more satisfying ultimately.

Some dogs are fed unlimited food which is low in nutrients. This allows the dog to determine its own food consumption. Some dogs do well on this type of diet, although those breeds that are predisposed to obesity, and certain other

individuals will eat the entire amount in one sitting if given the opportunity!

Foraging behaviour allows steady energy intake, and certain activities encourage this. Examples include hiding food, specific treat-obtaining toys, and distributing food on a tray or dish with a large surface area so that the food is slightly scattered and takes longer to eat. Havng to work a little for the food also means that your dog will be physically and mentally stimulated, adding interest to mealtimes.

Diseases and disorders

Insulinoma is a tumour that occurs in the pancreas and produces too much insulin, prompting increased appetite and therefore food intake, and promoting fatty tissue.

Pituitary gland deficiency gives rise to hypothyroidism; a common disease in dogs, caused by low levels of the hormone thyroid.

Cushing's syndrome or hyperadrenocorticism is a hormonal disorder that results from over-production of steroid hormones in the kidneys.

The hypothalamus is the part of the brain that controls appetite. When the hypothalamus is damaged, normal appetite control is compromised and overeating may result.

Environment and lifestyle (controlled exercise and regularity, everyday activities, sedentary behaviour)

Owners are responsible for exercising their dog on a regular basis; at least twice a day for the average canine. Working breeds, those with excess energy, and 'extra intelligent' dogs (Alaskan Malamutes, Siberian Huskies, Boxers, and Border Collies) need more exercise, ideally three times a day to ensure mental and physical stimulation is sufficient to preclude behavioural problems.

Inactivity cannot be compensated for by reducing food intake, and exercise alone will not result in the desired weight loss; it has to be a combination of both regimes.

WEIGHT MANAGEMENT

Since obesity has a major influence on canine health; management of the condition represents an important and frequent challenge for veterinarians and animal professionals. The use of portion control feeding, regular exercise, and avoidance of development of bad habits and human food should ensure a nutritionally balanced dog.

A change in diet should be made progressively for gradual but effective weight loss. To begin with, reduce food intake by between 20 and 40 per cent (the exact amount depends on the individual dog in respect of its starting and target weight). Weight loss can take months, and additional exercise should be introduced gradually to begin with. For instance, walks should be extended by

five minutes per walk each day, or taken at a slightly faster pace. Toys and games are a fun way to enhance activity as well. What activities you do with your dog – and for how long – depends on his age, health, and current weight.

Aquatic exercise can be very effective for weight loss as the increased resistance requires more energy input. Be aware, though, that the degree of obesity is important as complications that may compromise health are more likely to arise. A combination of land-based exercise and hydrotherapy will ensure an effective weight management programme, and help prevent further damage to the body. In this respect, perseverance is the keyword for owners with the promise of a healthier, happier dog at the end!

7
Hip dysplasia

SEVERE HIP DYSPLASIA

The femoral head is noticeably subluxated or luxated with obvious diminishing of socket and clear signs of osteoarthritis.

DIAGNOSIS

Observation, palpation and x-rays are necessary to determine hip dysplasia and its progression. There are several tests which vets can use, such as the Ortolani, plus the Bardens and Barlow tests (range of motion tests) to detect signs of osteoarthritis. Prior to any radiographic indications, structural changes occur to the muscles, ligaments, and cartilage. However, a dog that does not show any clinical signs of hip dysplasia may show dysplastic development when x-rayed.

SURGICAL MANAGEMENT

Surgery may be recommended depending on the stage of the condition; in some cases, an operation is not necessary, or may be considered at a later date, depending on how the condition develops. It is possible to manage and stabilise hip dysplasia with hydrotherapy and anti-inflammatory medication. Muscle development through exercise can provide enough support to hold the hip in the right position.

Several surgical procedures areavailable, including triple pelvic osteotomy (TPO), ostectomy, and total hip replacement (THR).

The TPO procedure is carried out in young dogs of between six and twelve months of age usually, with no or minimal signs of arthritis. Patients may show changes in gait, like the 'bunny hop' or lameness, be reluctant to jump, and move in an uneven way due to joint laxity, which can cause inflammation and abnormal positioning of the femoral head due to movement forces and weight-bearing. The reconstructive surgery involved in a TPO aims to allow joint congruency between the ball and socket in order for correct joint development to proceed without further damage. It does this by protecting the natural

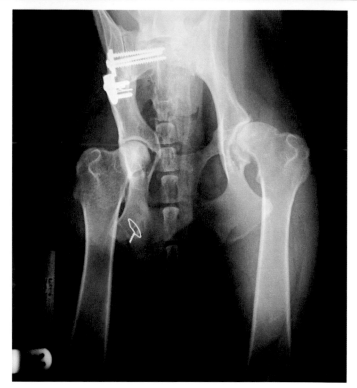

An x-ray of a German Shepherd taken four years after a triple pelvic osteotomy (TPO). The operation successfully prevented development of hip dysplasia.

hip joint, and eliminating subluxation and laxity, which helps prevent the progression of arthritis. The procedure involves cutting the pelvis in three places (sometimes four), and rotating the socket to provide better coverage of the femoral head, resulting in a stable hip joint.

Total hip replacement (otherwise known as total hip arthroplasty) is considered for dogs with severe hip dysplasia or chronic hip luxation. As the name suggests, the diseased ball and socket are replaced with prosthetic implants, providing patients with full hip function.

Femoral head and neck excision, or femoral head ostectomy, is regarded as the alternative and cheaper version of a total hip replacement, and is often carried out in small breeds because of fewer complications following surgery. An ostectomy is the surgical removal of part of a bone, and, unlike the total hip replacement, only the affected section of the hip is removed. A femoral head ostectomy is intended to prevent friction (bone to bone contact), thus causing less trauma and pain. In a dysplastic dog, the acetabulum changes shape and size to accommodate the contacting femoral head. After the procedure is complete, the femoral head is not replaced; instead, a false joint of scar tissue forms within the socket cavity. In addition, the muscles surrounding the hip area act as support.

All of the available treatments are costly, and may not prove completely effective in eliminating clinical signs preventing development of the condition. To create the best possible musculoskeletal environment for pain-free hip function, and to delay or prevent the onset of degenerative joint disease (osteoarthritis), physical treatments, preventative therapies and rehabilitation have potential for the non-surgical management of canine hip dysplasia patients.

NUTRITION AND EXERCISE

The incidence of canine hip dysplasia has been linked to greater body size and the rapid growth rate which usually occurs between four and ten months of age, so primarily affects medium, large and giant breeds, and those with high mature weight and muscle mass. Most commonly affected are German Shepherds, Golden Retrievers, Labrador Retrievers, Rottweilers, Bernese Mountain dogs and Spaniels.

Nutrition and exercise may alleviate or delay the onset of clinical signs and clinical symptoms. However, over-exercise and inappropriate feeding are likely to have the opposite effect. Restricted activity is recommended to avoid excessive wear on the affected joint, as well as to limit inflammation. Exercise should be short but frequent so as not to overload the joints. Hip dysplasia can be painful, so long bouts of exercise should be avoided for this reason also. Walking on grass, sand and soft surfaces will help cushion weight-bearing.

Jumping and walking up and down stairs should be kept to the absolute minimum, or ideally avoided, especially prior to adulthood.

In conjunction with controlled exercise, dogs should be fed strictly measured meals with limited treats and no human food. A variety of nutritional and mineral supplements (such as green-lipped mussel extract) have been used with varying degrees of success: consult a vet for advice on this subject.

Case history

Name: . . . . Alfie
Breed: . . . . Newfoundland
Age: 4
Sex: . . . . Male
Weight (at start): 69.2kg (10.9st)
Condition: . . . . Hip dysplasia
Owned by: . . . . Leanne Mundy
Surgical procedures: . . . . Total hip replacement
Medication: . . . . None

INTRODUCTION TO HYDROTHERAPY

Leanne first noticed that Alfie was limping in January 2009, when he was 3 years old. Alfie had not displayed any other clinical signs previous to this.

After a veterinary examination, hip dysplasia was not thought to be the problem, and Alfie was prescribed anti-inflammatory medication to see whether it would improve his lameness. There was no improvement, and so Leanne took Alfie back and radiographs were taken. These revealed severe bilateral hip dysplasia and Alfie was referred to a specialist where a total hip replacement was successfully performed on his left hip (the worst affected).

To ensure complete recovery, Leanne was advised to rest Alfie and limit his exercise, a consequence of which was that he was unable to burn calories from his normal food intake. Due to the added strain of helping to support the left hip, Alfie's right hip began to deteriorate. It is difficult for large breeds to weight-bear on three legs, and Alfie became uncomfortable and lame. In May 2010, a total hip replacement was carried out on his right hip.

The vet suggested hydrotherapy as an aid to recovery but did not refer Alfie. Leanne, as a (grooming) customer at Animal Magic, told staff there about Alfie's understandable inability to stand for long periods of time, and the need for extra consideration during the handling of his hind legs due to his condition, after which he was referred for hydrotherapy.

AIM OF PROGRAMME

To enable Alfie to carry out exercise in a safe environment, without fear of weight-bearing stresses and strain to his body, whilst improving flexibility, mobility and strength in order to support his weight.

PROGRAMME DURATION

Alfie began hydrotherapy on 26 August 2010, completing a four minute session.

TASKS TO BE COMPLETED AT THE CENTRE

Free swim in a buoyancy jacket with a hydrotherapist to encourage circular and figure of eight patterns. Tasks to be completed at home: Reduced food

Newfoundlands are renowned for their love of water.

Alfie started off as a very over-excited swimmer ...

intake, no jumping, no stair climbing or vigorous land-based exercise.

PROGRESS REPORT

Alfie had very high energy levels which were helped and managed by the introduction of tennis ball retrieval in the pool. Having him swim in figures of eight and in different directions encouraged resistance and strength, especially in his hind region which aided flexibility and strength.

Alfie had good body condition when he first attended Animal Magic, though would have benefited from losing some weight. Leanne has Alfie clipped in the warmer months of the year, which allowed a good assessment of his physique.

The average weight of a Newfoundland of Alfie's measurements is 68kg (10.7st). Subsequently, Leanne refined Alfie's diet, and with hydrotherapy, he managed to lose some weight, allowing him to be more mobile and energetic.

Since Alfie had undergone two major operations, and was unable to carry out serious exercise, his cardiovascular fitness level was below par. Alfie took to the water like a typical Newfoundland and thoroughly enjoyed his sessions. Unfortunately, a previous ear complaint meant that further sessions had to be postponed until it could be established whether or not the hydrotherapy was aggravating it.

Alfie attended once a week for sessions of up to 11 minutes. He had a total of ten sessions of water therapy, with his last swim on 28 October (hydrotherapy was discontinued due to the recurring ear condition). During this time, Alfie's muscles did begin to build and his general fitness improved greatly. His new tolerance to exercise kicked off his weight loss programme, and enabled him to support his body on land without lameness.

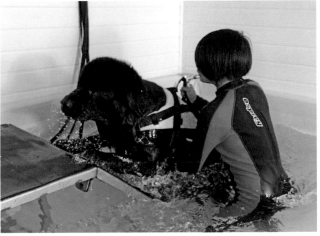

Ramp access is an important feature, especially with heavy breeds.

... but after his first few sessions, Alfie's strokes became much more relaxed and measured.

Case history

Name: . . . . .Wally
Breed: . . . . .Border Collie
Age: 1
Sex: . . . . .Male
Weight (at start): 23kg (3.62st)
Condition: . . . . .Bilateral hip dysplasia
Surgical procedures: . . .None
Medication: . . . . .Synoquin

INTRODUCTION TO HYDROTHERAPY

Graham and Susan took Wally into their home when he was 10 weeks old. When out walking one day, they noticed that Wally had developed a slight cough, and kept sitting down as if he was in discomfort. On a visit to the vet three days later, Wally was diagnosed with bilateral hip dysplasia.

The vet advised Graham and Susan to provide Wally with 15 minute, lead-restricted walks no more than three times a day in conjunction with hydrotherapy; surgery was not recommended. Wally was already attending a hydrotherapy centre but transferred to Animal Magic for treatment.

AIM OF PROGRAMME

Build muscles around the hip area to support the joints, whilst increasing strength and mobility to provide comfort and increased wellbeing. The main objective was to prevent the need for surgery.

Wally required encouragement to improve his technique, as he was quite nervous to begin with ...

PROGRAMME DURATION

Wally first swam at Animal Magic on July 9, 2010, and underwent hydrotherapy twice a week, reducing to a 20 minute session once a week from August 26.

TASKS TO BE COMPLETED AT THE CENTRE

Swim in a buoyancy jacket in the centre of the pool, rotating both clockwise and anti-clockwise, which ensures that Wally uses his hindlegs and lower back muscles to turn, strengthening and encouraging flexibility which on land will help stabilise him.

TASKS TO BE COMPLETED AT HOME

The Jarvis family was advised to prevent Wally jumping up or down, and walking up and down stairs; also to reduce ball games, restrict walks to on the lead, and monitor food intake to keep weight at an optimum level.

PROGRESS REPORT

Wally's energy levels have appeared to increase,

... although, because Wally adapted very quickly, it wasn't long before he was enjoying ball games!

Recommended exercises include work with a cavaletti (a ladder-like piece of equipment that is raised slightly off the ground, which dogs walk over), turning left and right, as well as executing figures of eight.

Wally's body shape is better defined, and his muscle tone has developed sufficiently to support his hips and carry his weight. Over the course of his hydrotherapy Graham and Susan have successfully maintained Wally's weight at 23kg.

Wally started a conservative management programme on 21 October, attending hydrotherapy sessions once every fortnight, and swims to maintain his condition and delay the need for surgery. Although Wally is only young, it is likely that he will develop osteoarthritis early.

Wally's hydrotherapy sessions have decreased subsequently, but his land-based activities have increased because he is physically fit enough. Wally has come on in leaps and bounds: without hydrotherapy, he would have quickly deteriorated.

A shower after a swim removes chlorine from a dog's coat, which, if left, can cause skin problems.

It's hoped that hydrotherapy sessions will mean that surgery will not be necessary.

and his exercise tolerance has improved. Wally now manages 20 minute sessions in the pool whilst catching tennis balls. At home, he can now play and retrieve toys, as well as go up and down stairs.

Wally attended several physiotherapy sessions and the physiotherapist was pleased with his progress.

Due to his previous experience and fitness level, he undertook an 8 minute program. He was scheduled for fortnightly sessions.

TASKS TO BE COMPLETED AT THE CENTRE
To swim with a buoyancy jacket attached to the rear 'D' ring, to offer extra support to the hip region until the muscles sufficiently developed for him to be able to independently hold his pelvic region correctly in the water.

TASKS TO BE COMPLETED AT HOME
Kevin and Tracy conducted and organised physiotherapy, massage, flexion and extension exercises, restricted walks, and limited activities.

PROGRESS REPORT
As a typical water-loving Labrador with previous swimmingexperience, Rory was an excellent swimmer. His swim duration built quickly, along with his body conditioning. As he was swimming every fortnight, it was important not to overwork him. Rory now has sessions of up to 13 minutes every fortnight for maintenance purposes.

Considering the trauma that his body has experienced, Rory has coped tremendously well; he's a true fighter. When Rory first attended hydrotherapy at Animal Magic, he was unstable in the hind region due to muscle wastage (atrophy). He now has the support of his muscles, and sufficient strength to enable him to play, and run in a controlled way without becoming fatigued and unsteady. With the care and attention of his owners, Rory's weight has been maintained and thus his body has not had to contend with extra pressure, enabling him to achieve and maintain a good level of fitness.

Pelvic rotation in the pool is still evident; this is when the pelvic region curves (toward the ribcage) slightly when swimming instead of staying relatively straight, in line with the rest of the body. This curving means that muscles around the spine and front of the body are needed to move the hind limbs in the water, as opposed to mainly using muscle structures around the hips. The surgical procedures Rory underwent have left him perfectly capable of functioning normally, but unable to flex and extend his hind limbs 'correctly' with a tendency to hyperextend.

Rory has been given a new lease of life, and, at the age of two, is still catching up with puppyhood. Hydrotherapy sessions for the rest of his life will mean that Rory will be able to maintain body condition, and help keep his left hip in check. Osteoarthritis is inevitable, and unfortunately is likely to occur at a young age; however, aquatic exercise will help to manage this.

Hydrotherapy is the safest way for Rory to exercise, and will help slow the progression of osteoarthritis.

Case history

Name: . . . . Ralph
Breed: . . . . Labradoodle
Age: 2
Sex: . . . . Male
Weight (at start): 30.3kg (4.77st)
Condition: . . . . Bilateral hip dysplasia
Owned by: . . . . Phil and Maggie Crathern
Surgical procedures: . . .None
Medication: . . . . .Previcox

INTRODUCTION TO HYDROTHERAPY

Maggie and Phil noticed that Ralph was lame in his hind legs, and seemed unstable. When he was five-and-a-half months old, Ralph was diagnosed with bilateral hip dysplasia, when an x-ray revealed severe right hip subluxation and left hip luxation. Because of his young age, surgery was not an option, and he was therefore prescribed Carprieve, a type of analgesic medication, to relieve the pain caused

by inflammation of the joints. Ralph experienced vomiting as a side effect of his medication, and was given Previcox as an alternative. It was hoped that the anti-inflammatory medication would minimise his clinical signs and discomfort, thus managing the condition and negating the need for corrective surgery. At the time of examination, Ralph was showing no signs of osteoarthritis; surgery was to be reconsidered in the future if arthritis was progressing, and if any deterioration was evident.

Ralph is a very active and playful dog, which, unfortunately, was not helping his condition: the vet advised that his exercise be restricted. This was a difficult proposition for Maggie and Phil as there were other dogs in the family, and Ralph became frustrated about not being able to interact with them. Even though he had very high energy levels, Ralph had evident muscle wastage around his pelvic region, which increased joint instability. As a Labradoodle, his growth development and predicted adult conformation meant that he was going to be a very large dog, his increasing weight putting more strain on the body. At the time of his examination, happily, Ralph was showing no sign of osteoarthritis.

It was vital that Ralph built his strength to reduce the likelihood of further problems associated with hip dysplasia and fast growth rates, and give him the best chance of a normal life. The vet suggested

Care must be taken during the growth period of large breeds to assist prevention of developmental disorders.

continued over

Phil with Ralph and companion Albert.

Ralph swims very steadily, and needs no encouragement.

hydrotherapy as a form of exercise not only for Ralph's condition, but also to burn off excess energy. Ralph was referred for treatment immediately.

AIM OF PROGRAMME
To build and maintain muscle mass to support the hip joints, manage any osteoarthritic changes, and obviate the need for surgery.

PROGRAMME DURATION
Ralph first came for hydrotherapy on 27 October 2009, and had five minute sessions, three times a week. Tasks to be completed at the centre: To swim in a buoyancy jacket with rope attachments to the rear 'D' ring for pelvic support. Due to his long ears and low head disposition, Ralph wore a head flotation ring to lift his head and keep it in line with the rest of his body (and also prevent his long ears from getting too wet). Tasks to be completed at home: Short lead walks only, no jumping or stair climbing, and strict weight management.

PROGRESS REPORT
Ralph's water therapy sessions were reduced to two times a week from 1 December 2009, as his fitness level

Rope attachments are used to support Ralph; he becomes panicked with a therapist in the pool.

was sufficient to enable him to have less frequent but more intense sessions.

Ralph has always been a relaxed and steady swimmer, taking every stroke at his own pace. To encourage more cardiovascular intensity, on 26 January 2010, the jets (which cause turbulence)were introduced. Subsequently, Phil and Maggie reported back to Animal Magic that Ralph had demonstrated some stiffness and lameness; a result of over-exercise. Consequently, from 30 January, the jets were no longer used and his sessions were reduced to one a week. Ralph coped well with longer swim durations (up to 20 minutes) at a lower intensity, which proved effective for stamina and strength development.

The vet has been pleased with Ralph's progress so far, although there is a possibility that a total hip replacement procedure will be required for both hips at some point. Ralph's weight has been maintained throughout his program, although it was advised that he should lose weight. Phil and Maggie have put him on a diet to reduce the weight that is putting pressure on his joints. Even though he is not considered overweight and is in good overall condition, a dog in his circumstances should be as lean and fit as possible.

On 20 November, unfortunately, Ralph ruptured his cranial cruciate ligament and so hydrotherapy was postponed until he had sufficiently recovered from a lateral imbrications procedure. Hydrotherapy will be beneficial for the dysplasia and rehabilitation of his left hind leg, assist in prevention of a secondary cruciate injury to his right hind leg, and help ward off osteoarthritis. A cruciate rupture will put added strain on Ralph's hips and further complications will occur if his right cruciate also becomes injured. Strict weight management and exercise restriction (no playing, reduced land-based activity), although appearing unfair, is vital to Ralph's recovery.

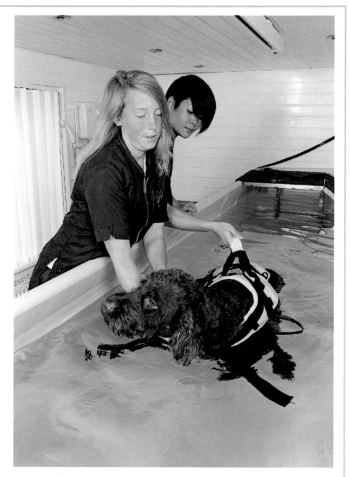

Ralph is lifted into and out of the pool as he is not keen on the ramp.

8
Elbow dysplasia

The elbow is made up of three bones: the humerus, radius, and ulna. The radius is located at the top of the forelimb, and connects with the humerus to form the carpus, or wrist joint, at the lower end of the forelimb. At its top end, the radius articulates and supports the humerus, which allows weight-bearing. The top of the ulna curves around the humerus to allow normal movement of the elbow joint with what is known as the trochlear notch; a concave indentation. In the normal elbow joint, the humerus fits neatly in the trochlear notch of the ulna to create a hinge-like joint. Although it is a separate bone, the ulna is closely attached to the radius by muscle.

Abnormal formation or dysplasia of the elbow can occur through genetic inheritance. A combination of genetics, breed and lifestyle contribute to its presence. Primarily, young, fast growing, and large breed puppies develop elbow dysplasia, although it can also be seen in small and adult dogs. Overfeeding protein and high calorie diets – along with over-supplementation of vitamins and minerals (especially calcium) – can worsen or accelerate development in fast growing puppies. All breeds can be affected, with chondrodystrophic breeds (those bred to have crooked legs, such as the Bassett Hound and Staffordshire Bull Terrier) commonly experiencing abnormal development of the elbow joint due to atypical weight-bearing pressure.

Elbow dysplasia incorporates several diseases; ununited anconeal process, fragmented coronoid process and osteochondrosis/osteochondritis dissecans. Commonly, more than one condition will be present in the elbow, occurring both singly and bilaterally.

Approximately 60 per cent of a dog's body weight is carried on its front legs, so even if suffering from canine hip dysplasia, for example, he can cope if the muscle structures in the pelvis area are strong enough to stabilise and support his weight. However, a heavy load on elbow joints through compensatory mechanism predisposes a dog to mild elbow dysplasia. Compensatory mechanism is when the body relies more heavily

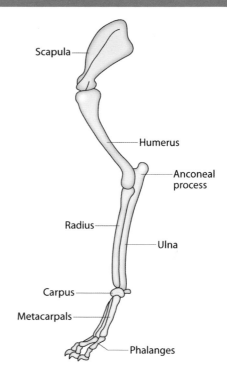

As the forelimbs support 60 per cent of a dog's body weight, any abnormalities will further stress the joints. (Courtesy Dave Russell)

on a limb or area to make up for the loss or dysfunction of another limb or area, causing it extra/abnormal effort. Once elbow function is compromised, foreleg gait becomes affected as greater strain is placed on the other leg. Once abnormal development has begun with what is known as a primary lesion, further abnormal wear of the joint surfaces and secondary arthritis will occur. Secondary arthritis cannot be reversed and remains a potential problem for the rest of the dog's life.

DIAGNOSIS
Elbow dysplasia can be diagnosed by radiographs,

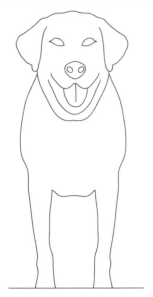

Most breeds show 'normal' straight leg conformation. (Courtesy Dave Russell)

CT scans, palpation, and assessment involving motion of the foreleg. Heat and swelling from inflammation may be evident, as well as some pain responses. Diagnosis will help determine the exact type of condition involved, and allow prescription of the most appropriate treatment. For most dogs, the best treatment is conservative management involving changes in diet and exercise, and modification of day-to-day activities and tasks, although surgery is most effective in cases where the elbow dysplasia has caused arthritic changes in the joint. The procedures are relative to the condition and stage of development.

PREVENTION

Like the prevention scheme for hip dysplasia, the canine health scheme developed a grading system to detect and help prevent the genetic factors of elbow dysplasia from being passed on.

Radiographs are taken of both forelimbs, and examined independently. To ensure that those areas of the joint where abnormalities occur can be examined, two x-rays of different angles are taken. Primary lesions and arthritic changes are observed, and each elbow is then graded, the overall grade determined by the higher of the two. For example, a left elbow score of 2 and right elbow score of 0 will mean that the dog has a score of 2; elbow dysplasia is graded 0-3, 0 being normal and 3 severe. If a dog owner or breeder wishes to, they can have their dog examined for elbow and hip dysplasia at a discounted price! Again, this scheme relies on voluntary submission for analysis at the owner's cost.

● No signs of elbow dysplasia
There is sound conformation of the elbow joint. The humerus, radius and ulna are correctly formed and congruent. No signs of osteoarthritis

● Mild elbow dysplasia
Slight primary lesions are present. Primary lesions relate to incongruency, misalignment of bones, and mild deterioration of the trochlear notch. No signs of osteoarthritis

● Moderate elbow dysplasia
More prominent primary lesions can be seen, including fragmentation and united processes, as well as the onset of osteoarthritis

● Severe elbow dysplasia
Primary lesions are obvious with clear signs of secondary osteoarthritis.

UNUNITED ANCONEAL PROCESS

Ununited anconeal process is a failure of the growth centre of the anconeal process in the elbow joint to unite properly with the ulna (this fusion should be completed by sixteen to twenty-four weeks of age). Instead of a normal bony union, the ununited anconeal process represents a large piece of bone connected to the ulna by a strand of fibrous tissue. This incongruity between the rate of growth of the radius and the ulna results in pressure on the point of the anconeal process, causing pain and inflammation. Consequently,

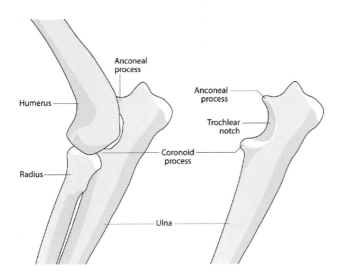

An ill-fitting elbow joint can cause lameness.
(Courtesy Dave Russell)

management is not proving effective; however, osteoarthritis may be too advanced for a truly effective outcome.

FRAGMENTED CORONOID PROCESS

Fragmented coronoid process is the most common form of elbow dysplasia caused by incongruity radius and ulna lengths. As a result, the coronoid process (a triangular eminence projecting forward from the upper and front part of the ulna) can become disturbed and fragmented. Fragmentation occurs when parts of the bone and cartilage become detached and break off from their usual position. The fragments rub and irritate the cartilage surfaces, causing additional joint damage and the early onset of osteoarthritis.

Clinical signs are rarely noticeable before a dog is five months of age. Slight lameness is common with early cartilage damage. Unfortunately, because cartilage material is not dense enough to be detected clearly on an x-ray, it cannot be used for diagnosis in the early stages of the condition. In later development, the beginnings of osteoarthritis associated with fragmented coronoid process can be seen with an x-ray. Alternatively or additionally, a CT scan may show incongruity not detected with a radiograph.

Fragmented coronoid process is often seen in dogs with ununited anconeal process, and is thought to have derived from osteochondritis dissecans (see below). Fragmented coronoid process is usually treated surgically by a fragment removal procedure inside the elbow joint. The fragments are eliminated to help avoid severe osteoarthritis (although osteoarthritis is still expected to develop).

OSTEOCHONDROSIS/OSTEOCHONDRITIS DISSECANS

Osteochondritis dissecans mainly affects medium, large, and giant breeds, and is strongly associated with the genetic factors responsible for weight gain, growth, and conformation. Osteochondritis dissecans is a developmental abnormality in the

an affected dog may show varying degrees of lameness between five and nine months of age, although this condition can also be found in older dogs.

Ununited anconeal process is typically found in large and giant breeds, with male dogs being more susceptible than females, which suggests that the condition has genetic elements. Ununited anconeal process also occurs in chondrodystrophic (congenital dwarfism) breeds as a result of retarded ulna growth giving rise to elbow incongruency.

If ununited anconeal process is identified early enough (four to five months of age), corrective surgery is possible by cutting the ulna to improve the shape of the elbow joint and make it more congruent. An ulna osteotomy or radial lengthening osteotomy may be carried out to cut and equalise radius with ulna lengths. This is often carried out in adult dogs where conservative cartilage of puppies, which results in areas of bone where the cartilage is thicker than normal. Osteochondritis dissecans may also be present in the stifle, shoulder and hock – all structures that experience increased day-to-day weight-bearing loads – and gives rise to the formation of a cartilage flap or lesion, as well as free-floating fragments in the joint, which cause pain and inflammation. Pain, limping lameness and stiffness are usually demonstrated at six months of age. Surgical removal of the cartilage lesion is required to eliminate pain, and will encourage new, healthy cartilage formation which takes about three weeks to grow (but eight weeks to stabilise enough to withstand everyday pressures).

Case history

Name: . . . . Conker
Breed: . . . . Labrador Retriever
Age: 2
Sex: . . . . Male
Weight (at start): 28kg (4.4st)
Condition: . . . . Elbow dysplasia
Owned by: . . . . Adam Jupp and Leanne Deacon
Surgical procedures: . . . Triple pelvic osteotomy for previous hip dysplasia
Medication: . . . . Cartrophen, Previcox

INTRODUCTION TO HYDROTHERAPY

Conker became a part of Adam's and Leanne's lives at 8 weeks of age. He was bought from a private breeder where good hip scores of the sire and dam were submitted for analysis. When he was eleven months old, Conker was diagnosed with dysplasia of his left hip. As a result, a successful triple pelvic osteotomy was performed. After a long recovery process (without hydrotherapy), Conker was almost back to his former self. However, as predicted with all post-surgery patients; further complications had begun to occur.

In June 2010, Conker began to demonstrate lameness in his right foreleg. As dysplasia was present in his left hip, it is likely that Conker compensated for this by shifting his weight to his front legs to alleviate stress and discomfort. Conker was taken to the vet for an investigation, where a radiograph revealed the onset of arthritis in his elbow.

Conker has always suffered with problems associated with his elbows due to conformation abnormalities, which cause his elbows to point inward. Given the abnormal forces on the elbows throughout his growth development as a result of this, and associated osteoarthritis, it was likely that Conker had undiagnosed elbow dysplasia. It could have been the hip dysplasia and abnormal elbow conformation that caused the arthritis, as well as the elbow dysplasia. To restore fluids and provide lubrication to reduce friction of the joint surfaces, he was given a course of cartrophen injections and referred for hydrotherapy.

AIM OF PROGRAMME

To allow Conker a form of exercise where his joints are protected from constant land-based stresses that would complicate his condition. Exercise goals were to maintain the elbow dysplasia, and improve flexibility and function.

Conker shows no clinical signs of dysplasia or arthritis.

PROGRAMME DURATION
Conker first came to Animal Magic on 15 July 2010, attending three times a week in sessions developing from 4 to 12 minutes.

TASKS TO BE COMPLETED AT THE CENTRE
Conker was to swim in a buoyancy jacket with rope attachments that allowed the therapist control of his movements. He is strongly motivated by toys, and thoroughly enjoys retrieving rubber ducks (in fact, he is the only dog who likes Animal Magic's ducks!). Retrieving toys encourages him to turn both clockwise and anti-clockwise, using all his limbs, whilst maintaining control with the rope attachments.

TASKS TO BE COMPLETED AT HOME
Maintaining body condition (which has always been ideal), allow gentle exercise, limited jumping.

PROGRESS REPORT
When Conker first came for hydrotherapy, he began with sessions of 4½ minutes' duration. Conker became tired quickly when playing with and chasing the toy ducks, but his flexibility and mobility have improved through hydrotherapy; this is evident in his gait on land. Without knowing the details of Conker's condition, it could easily be possible to believe that there was absolutely nothing wrong with his elbow.

Conker is now able to remain active for long periods of time without demonstrating lameness. Consequently, his fitness and stability improved to the extent that, from 24 August 2010, his sessions were reduced to twice a week. A month or so later, his hydrotherapy sessions were reduced again as his land-based activity was slowly increasing, and he

Conker's condition has not affected his activity levels or abilities: he loves being outside.

continued over

attended just once a week, for up to 18 minutes each session.

When re-examined by his vet, Conker had progressed so well that, from 16 October, his sessions were reduced to once every fortnight for fun, fitness and maintenance. Since December 2010, Conker's programme has become more challenging and he is asked to free swim whilst retrieving in patterns of circles, figures of 8, and zigzags. These exercises require more strength and stamina, and Conker has reduced sessions of 15 minutes which are of higher intensity.

Conker is still young but such an early diagnosis of arthritis means that it is likely to develop and affect him more rapidly in later life. Hydrotherapy and ongoing weight management will minimise and help slow development of the degenerative joint disease in both his hip and elbow, although additional stress is likely to cause osteoarthritis to develop in both his other limbs, so conservative management is vital in maintaining his quality of life.

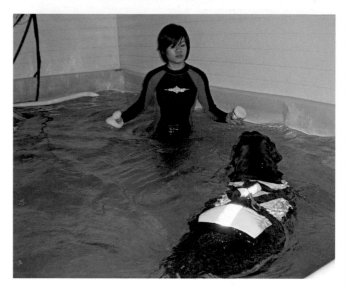

Conker now enjoys fun and fitness swims to maintain his health.

Retrieving ducks encourages him to flex and extend his legs thoroughly.

Rope attachments were first used to control his movement.

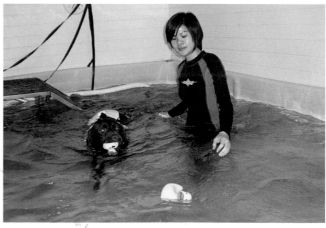

9
The patella (knee cap)

The patella (knee cap) is a flat, triangular bone located at the front of the stifle (knee) joint, and connects the femur to the tibia. There are two joints in the knee that together work like a hinge to allow the knee to bend, straighten, and rotate slightly from side-to-side. The knee cap aims to relieve friction between the femur, tibia and muscles, as well as help protect the joint. The knee cap normally sits in the trochlear groove within the stifle. The femoral groove (as it is also known) is a concave surface where the patella makes contact with the femur.

Four ligaments hold the knee in place and control stability. A pair of collateral (parallel) ligaments prevent the knee moving too far sideways, and the cranial and caudal cruciate ligaments interweave in the centre of the knee, and prevent too much forward and backward movement. Additionally, the patella ligament attaches the knee cap to the tibia and assists with support. Extra stability is provided by two major tendons: the quadriceps and patella. The quadriceps provides strength for straightening the knee as well as retaining the knee cap in the trochlear groove, whereas the patella tendon simply connects the patella to the tibia.

Patella luxation, or dislocation of the knee, is a condition where the patella sits outside the femoral groove when the knee is flexed (bent), but maintains tension as the knee is extended (straightened). It can be classified as medial or lateral, depending on whether the knee cap rides on the inner or outer part of the knee. Every time the knee cap rides out of its groove, cartilage is damaged, leading to osteoarthritis and associated pain. Deformation caused by breed conformation and growth development of the leg may cause abnormal alignment of the knee cap, which may also increase the depth of the femoral groove. Incorrect positioning of the knee changes the joint structure of the knee, which can predispose affected dogs to a rupture of the cruciate ligament. Similarly, a dog with cruciate ligament injury (see *Further reading*) can develop the later onset of patella luxation. Traumatic injury to the

Hind leg deformation caused by breed conformation and growth development can make a dog more susceptible to patella luxation. (Courtesy Dave Russell)

Patella luxation and cruciate injury are interlinked conditions, as they both occur within the stifle (the knee-like joint above the hock in the hind leg). (Courtesy Dave Russell)

knee causes acute non-weight-bearing lameness of the limb. In cases of non-traumatic patella luxation, the trochlear groove is commonly shallow or deficient.

Bilateral luxation (in both knees) is typical, although unilateral (one knee) luxation does occur. Patella luxation has been linked to genetic and developmental factors, with most breeds being susceptible. Small breeds, like terriers, are most commonly predisposed to the condition, the majority affected by medial patella luxation (when the patella slips out of place and goes to the inside of the leg). Larger breeds are infrequently affected, but when they are, lateral patella luxation (when the patella slips out of place and goes to the outside of the leg) is usually diagnosed.

Dogs of all ages can develop patella luxation. The condition can be caused by abnormal conformation of the hip joint due to hip dysplasia, for instance, malformation of the femur and/ or tibia, and deviation of the tibial crest (the bony prominence onto which the patella tendon attaches below the knee).

Clinical signs vary from intermittent skipping gait to constant lameness, which may deteriorate or appear to right itself. Severely bilaterally affected dogs may be completely unable to extend the knee, subsequently adopting a crouching stance with limited movement.

Diagnosis of luxation is made by attempting to push the patella out of the trochlear groove. The degree of luxation is graded between 1 and 4, depending on how easy it is to dislocate the patella and whether or not the patella returns spontaneously to the trochlear groove. The table below shows the grades of severity and the definition for each grade.

GRADE 0
The patella is deemed normal and will not luxate during physical examination

GRADE 1
The knee cap will luxate when pressure is applied, but will return to its usual position when the pressure is released. A dog with grade 1 luxation is rarely lame, although it may occasionally skip (a skipping gait results from the knee cap slipping over the trochlear ridge)

GRADE 2
The knee cap can be manually luxated, and also during flexion and extension of the knee, returning to its usual position slowly. Lameness varies from a skipping gait to continuous lameness with the dog showing slight bow-legged conformation

GRADE 3
The knee cap will remain luxated most of the time, although it can be returned temporarily with manipulation. Lameness varies with grade 3 luxation from moderate to severe, together with associative bow-legged conformation

GRADE 4
The knee cap is constantly luxated and cannot be returned to its usual position in the trochlear groove (which is either too shallow or missing all together). Consequently, constant lameness is often seen, along with the inability to extend the knee, resulting in conformational deformity of the hind legs

DIAGNOSIS AND MANAGEMENT
Patella luxation diagnosis is made by feeling the knee area to detect any instability and assess ligament damage. Additionally, radiographs of the pelvis and knee are taken to determine the shape of the bones in the hindlegs, and rule out other possible conditions, such as hip dysplasia and bone degeneration (arthritis). X-rays will also enable a more accurate assessment of possible future arthritic conditions, and determine the best treatment.

Surgery is usually only considered for dogs that score grade 2 and above. A high percentage of dogs with patella luxation may show few or no clinical signs, and are prescribed conservative management to maintain the condition. However, dogs that become more lame and suffer deterioration of the condition may be reconsidered.

Surgical techniques can be categorised into those that involve bone structures and those that involve only soft tissue; most patients will undergo a combination of both procedures. Surgery aims to deepen the trochlear groove, repair the knee, and reconstruct any surrounding soft tissue to ensure it's correctly located. The type of surgery required is dependant on the grade of luxation, and what requires improvement: deepening the trochlear groove, realigning or repairing the knee, and/or tendons, ligaments and muscle structures to improve stability. Several surgical procedures are possible, as follows –

Trochleoplasty involves removing bone material from the trochlear groove to deepen it so that the knee cap fits within it correctly. As a result of the procedure, scar tissue grows to fill in the area where material has been removed, and holds the knee joint in place.

This procedure often causes complications, however, as bone is left in contact with the knee cap, resulting in friction and damage. To preserve contact between the cartilage and knee cap, improved procedures have been developed to prevent friction. These techniques are called the (elliptical) wedge recession and (rectangular) block recession method, in which a wedge or block shape of bone is cut and removed from the trochlear groove. Bone material is removed from the hole to deepen the groove, and the wedge or block then replaced in the newly-deepened groove, held in place by the pressure applied by the knee. With these techniques, scar tissue is not required to fill in the groove, which means that recovery is quicker.

The point at which the knee cap attaches to the knee ligament is called the tibial tuberosity, a bony area which often forms abnormally and becomes misaligned with the stifle joint. Realignment is necessary and this is done by tibial tuberosity transposition (TTT), which involves removal of the knee tendon and some of the tibial tuberosity, which are then reattached in a position that enables restoration of alignment of the quadriceps muscles, knee cap, and tendon.

To improve stability of the knee cap in position to prevent luxation, the imbrication technique – which tightens the joint capsule on the opposite side of the luxation – can be done. So, a dog with medial patella luxation undergoes a lateral imbrication, and a dog with lateral patella luxation undergoes a medial imbrication. Furthermore, the joint capsule can be loosened on the side of the luxation, which involves a retinacular release incision procedure (incision of tissue) to relieve tension on the knee cap, thereby enabling it to ride the trochlear groove properly. In some cases, osteotomy (where the actual bone is cut) is required in cases where the knee cap rides outside the trochlear groove most of the time.

Surgery is usually successful, with the majority of dogs recovering completely. Some lameness and stiffness may occur, but this is probably related to arthritic changes in the knee. In many cases, the incidence of patella luxation, especially in congenitally affected dogs, cannot be prevented, with most cases going undiagnosed. It is important for owners to take into consideration the possibility of patella luxation, as dogs that go undiagnosed and untreated are much more likely to develop further issues in the same or other hind leg. As the main weight-bearing joint, the consequences of problems with the knee could adversely affect a dog's long-term health and well-being quite significantly.

Case history

Name: . . . . Archie
Breed: . . . . West Highland Terrier x
Norfolk Terrier
Age:. 4
Sex:. . . . . Male
Weight (at start): 8.5kg (1.34st)
Condition: . . . . Patella luxation
Owned by:. . . . . Steve and Tricia Brooks
Surgical procedures: . ..Repair of iliac wings
and femoral shafts,
bilateral hip luxation
corrective surgery
Medication: . . . . None

INTRODUCTION TO HYDROTHERAPY

In 2007, Archie was involved in a serious road traffic accident resulting in trauma to the pelvis. Radiographs showed numerous fractures to his pelvis, and luxation of his left hip. Consequently, he underwent surgery to repair the ilium (top of the pelvis) and femur, as well as put his left hip back in place. Archie appeared to be recovering well at home, and hopes were high that he would soon be back to his usual, active self.

Archie was taken to the vet for a post-operative examination. His fractures were healing as was hoped, but much to the disappointment of Steve and Tricia, an x-ray revealed bilateral hip luxation. As a result, Archie underwent further surgery to stabilise both hips. Hydrotherapy was not used to aid his recovery.

At the beginning of 2010, Archie suffered patella luxation of the left knee, which seemed to self-correct. Hydrotherapy was then recommended to strengthen the supporting tissues around the knee in an attempt to maintain the patella luxation, and prevent the need for corrective surgery.

Archie had noticeable muscle atrophy, which it was hoped that hydrotherapy would help with, as well as develop and improve Archie's overall body condition.

AIM OF PROGRAMME

Water therapy to build muscle, improve flexibility, and develop fitness.

PROGRAMME DURATION

Archie's first session was on 24 June 2010, for 4 minutes in the pool.

Archie recovered exceptionally well from the accident.

continued over

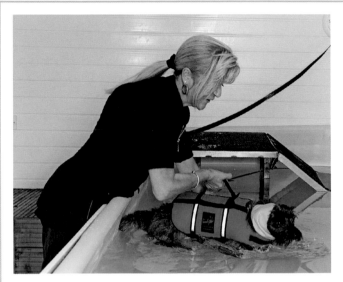

When Archie attended his hydrotherapy sessions, he was manually lifted into and out of the pool.

TASKS TO BE COMPLETED AT THE CENTRE

Swim with rope attachments in buoyancy jacket. To enable a neutral 'holding point' for balanced assistance from the buoyancy jacket, the attachments were added to the middle of the jacket. To prevent water entering his low-set ears, Archie wore a specially-designed head band.

Regular rests are important, especially when anti-swim jets are used.

Strategically angling Archie in the jets encouraged movement of his limbs. He wears a specially-designed swimming cap to prevent water entering his ears.

TASKS TO BE COMPLETED AT HOME

No jumping, and restricted exercise only to prevent further trauma to the knee and hips.

PROGRESS REPORT

After Archie's first few sessions, it was noticed that he needed a head band-type contraption to prevent water from entering his ears. By 6 July, Archie was managing hydrotherapy sessions of up to ten minutes, and his fitness levels developed quickly.

He continued to complete sessions of up to 15 minutes with rest, but required higher intensity exercise to get him working harder and stimulate him further. Increasing resistance in the water by using the jets was tried, and Archie seemed more content with this programme, which started on 3 August. To prevent Archie from becoming over-worked during these jet-assisted sessions, his 15 minute programme was divided into ten minutes without jets and five minutes with jets.

On 23 September 2010, Steve and Tricia stopped Archie's hydrotherapy sessions, due to a change in personal circumstances, but are intending to resume them in the near future for maintenance and 'fun' swims. Archie's energy and activity levels reportedly increased after his sessions, his mobility increased, and his knee stabilised. Muscle development and non-weight-bearing exercise via hydrotherapy has helped strengthen his cruciate ligaments to prevent cruciate injury, a common condition linked with patella luxation. Hydrotherapy, if carried out on a regular basis on a long-term scale, will enable Archie to maintain his knee in an attempt to prevent further luxations, arthritis, and corrective surgery.

Visit Hubble and Hattie on the web: www.hubbleandhattie.com and www.hubblendhattie.blogspot.com
Info about all H&H books • New book news • Special offers

83

10
Cruciate injuries

Several structures make up the knee joint, including the caudal and cranial cruciate ligaments which cross over one another inside the joint. These strong, fibrous bands join the femur (thigh bone) to the tibia (shin bone), and together work like a hinge, allowing the knee to bend whilst providing stability. A cruciate rupture of the cranial ligament is one of the most common orthopaedic conditions, resulting in hind limb lameness and osteoarthritis of the knee; however, it is rare for the caudal cruciate ligament to become injured.

The shin bone is held in place by the cruciate ligament (caudal and cranial). Following a rupture, the knee becomes unstable when weight is applied to it, because the top of the shin bone is no longer perpendicular to the length of the thigh bone. Resulting instability and pain mean that the dog will tend to hold up the affected leg.

Cruciate ligament injury can occur acutely because of trauma (a fall, knock or jump), or chronically (over time). Tearing of the cruciate ligament causes instability of the knee joint, which ceases to function properly. Most cruciate ligament tears in dogs occur gradually, resulting in low-level lameness that may or may not improve over time. Cruciate injury in a chronic form involves slow, progressive degeneration, which can lead to a partial tear or complete rupture (the latter exacerbated by trauma, or actually caused by trauma, depending on the strength and condition of the ligaments. However, it is common for the injury to occur without any particular incident and appear 'random.'

While acute ligament injury may occur because of a single incidence of trauma, the majority of ruptures occur during ordinary daily activity, due to secondary progressive and irreversible degenerative changes within the ligament itself.

Precisely why the cruciate ligament ruptures is not entirely understood, but is believed to be due to a combination of bone development, body conformation, and subsequent gait abnormality. Abnormal bone growth may result in irregular pressure and forces on the ligaments that lead to the degeneration process. Cruciate injury occurs

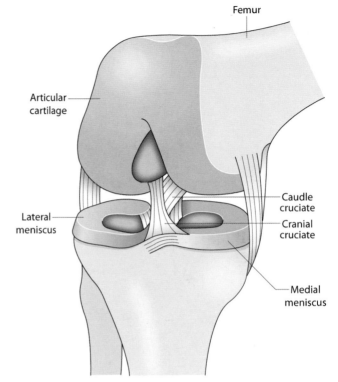

It is more common for the cranial ligament to rupture than the caudal ligament. (Courtesy Dave Russell)

frequently in overweight, middleaged/older dogs, with larger breeds most commonly affected. All dogs can suffer from a cruciate injury, although some breeds seem more predisposed than others, which indicates the influence of a genetic factor. A dog that is over-exercised and/or overweight will put extra and unwanted strain on the knees as well as other body joints.

CLINICAL SIGNS

A dog with cruciate degeneration may show mild clinical signs which appear to resolve and then reappear. This is common in highly active dogs,

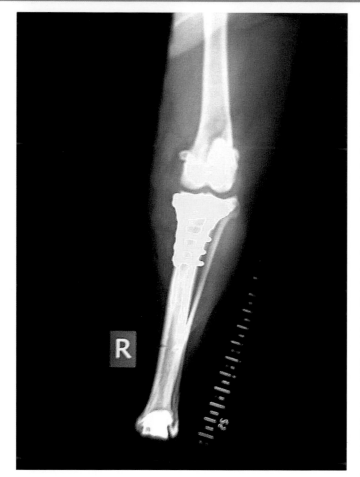

An x-ray after a successful tibial plateau levelling osteotomy.

and pain on palpation are also signs of injury. With some, a clicking or popping noise in the knee when walking may be heard; this is caused by meniscal damage. Menisci are two cartilage pads within the knee which act as a cushion when forces are applied to the kneww (eg weight-bearing), providing stability, reducing friction, and protecting articular cartilage. Meniscal incongruity prevents the meniscus distributing the forces transmitted through the knee joint, and, as a result, damages the articular cartilage of the joint, causing pain, lameness, and further rapid progression of osteoarthritis.

DIAGNOSIS AND SURGICAL TREATMENT

To diagnose cruciate rupture, a technique called the positive cranial drawer sign is used. The dog's knee is slightly bent and specific pressure applied. Subsequent sliding of the femur over the tibia (positive drawer sign) indicates cruciate ligament rupture. Often, with a chronic injury, the cranial drawer sign technique is less effective due to joint stabilisation resulting from a build-up of scar tissue in the joint capsule.

After the ligament tears, inflammation occurs within the joint. Although ligaments cannot be detected via radiograph, joint swelling can be seen; palpation to detect heat and inflammation can confirm this.

Surgery is often required to treat cruciate injury in order to stabilise the thigh and shin bone. Other objectives involve resolving knee joint instability, and managing any co-existing meniscal injury. Cruciate surgical techniques are categorised into intracapsular and extracapsular stabilisation: intracapsular relates to a procedure within the joint capsule, whereas extracapsular involves work outside and around the knee joint. Surgery should eliminate pain, resolve lameness, and reduce the rate at which cartilage is destroyed and arthritis progresses. The surgical procedure performed is dependent on injury severity, which ligament is involved, and whether it is a partial or complete rupture.

who may show pain and discomfort following exercise, coupled with difficulty getting up and jumping, which then appears to subside. Acutely affected dogs are often completely non-weight-bearing on the affected leg.

Depending on the progression of the degeneration, a dog will exhibit anything from mild intermittent lameness to full non-weight-bearing. Stiffness, muscle atrophy, swelling of the knee joint,

Intracapsular methods involve replacing the torn ligament. The most common cruciate surgery, especially in large breeds, is the tibial plateau levelling osteotomy. A tibial plateau levelling osteotomy involves changing the angle of the tibeal plateau by cutting the top of the tibia (osteotomy), rotating it, and stabilising it in a new position with plates and screws.

The extracapsular technique – performed on a dog with entire rupture of the cruciate ligament – does not involve replacing the torn ligament, instead, restoring the function of the original ligament. This procedure goes under several names, including extracapsular repair, lateral suture technique, and lateral imbrication technique.

The procedure involves special sutures being placed around the outside of the knee joint, anchored at the back of the thigh bone, and passed through a hole that is drilled in the shin bone. The sutures tighten the joint to hold it in place. As part of the recovery process, the body begins to develop scar tissue around the sutures at the knee joint, which acts to stabilise and hold the joint in place of the suture which will eventually stretch and/or break.

All surgical techniques have associated complications, including infection, patella luxation, screw loosening, fracture, and inflammation of the tibia, fibula and patella, limb oedema (swelling), neurological deficits, and late meniscal injury. Late menical injury occurs subsequent to cruciate surgery where the meniscus was correctly assessed to be normal, or where a meniscal injury was appropriately treated. A dog that ruptures one cruciate ligament has around a fifty per cent chance of rupturing other ligaments in the same and/or other knee.

Rehabilitation following surgery has been shown to improve muscle mass and improve muscle atrophy that occurs in the post-operative period, as well as increase knee range of motion, and recover weight-bearing ability. A working, competition or agility dog that has had corrective surgery will more than likely regain its former fitness and ability, although only if activity is drastically reduced during a long and complete recovery period.

Case history

Name:. . . . .Lilly
Breed: . . . .Yorkshire Terrier
Age:. 8
Sex:. . . . .Female
Weight (at start): 2.6kg (0.4st)
Condition: . . . .Anterior cruciate
                    ligament injury caused
                    by patella luxation
Owned by:. . . . .Mary and Lily Paul
Surgical procedures: . ..Anterior cruciate
                    ligament repair
Medication: . . . .None

continued over

INTRODUCTION TO HYDROTHERAPY

Lily and Mary took Lilly to be examined by the vet because she was walking on three legs only, and was reluctant to use and put weight on her left hind limb. An x-ray revealed a shallow trochlear groove; patella luxation. On examination, it was discovered that Lilly also had a cruciate ligament injury. Even though the patella luxation occurred first (which then caused the ligament injury), surgery was advised to replace and stabilise the anterior cruciate ligament. Specific surgical procedures for patella luxation cases in conjunction with cruciate ligament surgery have complications, and a high failure rate, so only the cruciate ligament was operated on.

After the operation, Lilly was still reluctant to use her left hindleg when walking, and she began to rely on her other legs; a common occurrence with small, lightweight dogs as they are easily able to distribute and balance their weight on three legs. Often, an injury which causes pain and discomfort can affect proprioception, an automated sensitivity/spatial awareness mechanism that sends messages via the central nervous system. The central nervous system then transmits information to the rest of the body about how to react and with what amount of tension. In Lilly's case, if bearing weight on her left hindleg resulted in repeated nerve pain, her body would quickly learn not to put the foot down. Even after surgery, Lilly still experienced heightened sensitivity, and needed to learn that weight-bearing would no longer be painful.

Lilly and her companion, Charlie, were taken to Animal Magic for grooming, where they were advised that hydrotherapy would help Lilly regain full use of her recovering leg. Lilly and Charlie, also a Yorkshire Terrier, are rescue dogs with a history of cruelty inflicted by previous owners. Mary and Lily wanted Lilly to have a good quality of life, and were willing to consider anything that would benefit their beloved companion.

During post-operative recovery, Lilly suffered a lot of muscle atrophy; although hydrotherapy was not initially recommended to Mary and Lily, their vet agreed that it would aid Lilly's recuperation and encourage her to use the bad leg.

AIM OF PROGRAMME

To increase mobility and flexibility of Lilly's left hindleg, assist muscle development, help relieve any pain, and

Cruciate injury frequently occurs subsequent to patella luxation, more so in miniature breeds. As a rescue dog, Lilly may have suffered trauma to her stifle.

help with arthritis management. As Lilly was not using her leg when walking, land-based activity was proving ineffective, so the aim of the hydrotherapy was to provide fitness exercise and improve body awareness.

PROGRAMME DURATION

Lilly suffers from laryngeal paralysis which limits the amount of endurance exercise she can do, as her respiration is affected (see progress report for full explanation).

Lilly began with sessions of one minute's duration with rest periods in-between, once a week from 29 July 2010.

TASKS TO BE COMPLETED AT THE CENTRE

Lilly began swimming, assisted by a hydrotherapist for the first few sessions, and was then allowed to free swim in a buoyancy jacket with plenty of rests. A typical session would involve Lilly swimming for 2 to 3 minutes, up to 3 times with rests in-between.

TASKS TO BE COMPLETED AT HOME

No jumping on and off furniture or travelling up and down stairs.

PROGRESS REPORT

As Lilly suffers from laryngeal paralysis, it was important to carefully monitor her breathing. Laryngeal paralysis is a condition where the muscles and nerves of the larynx do not open and close properly. The larynx is part of the respiratory system located at the back of the throat, which enables air to enter the body via the trachea (windpipe). Laryngeal paralysis can cause breathing to become laboured, preventing air from entering the body at the rate it requires to function properly. When dogs are stressed or panicked, breathing rate increases and the larynx is under more pressure to function. Care must be taken with dogs suffering from this disorder being able to breathe properly is obviously important, especially during a cardiovascular exercise like swimming. Serious dyspnoea (difficulty breathing) can result in death due to lack of oxygen.

Lilly had never swum before, and although dog's can swim, the experience was somewhat alien to her. At the very beginning, hydrotherapy sessions were of just one to two minutes' duration with plenty of rest. As Lilly is very small, when she swims her body is barely under the surface, where the water is a lot more viscous and thus more resistant to movement through

Lilly is the smallest dog to swim at Animal Magic to date.

it, making swimming for Lilly a lot harder than it looks.

After six weeks, Lily and Mary could see a marked improvement in Lilly's walking, energy levels, and activity. Lilly continued hydrotherapy on a weekly basis until 21 October 2010, at which point her swim duration had increased to 3 minute sessions. Mary and Lily decided that they would therefore stop her sessions as the main objectives had been met and everyone at the centre was pleased with her progress.

Case history

Name:. . . . . .Domino
Breed:. . . . . .English Springer Spaniel
Age:. 7
Sex:. . . . . .Male
Weight (at start):. 19.3kg (3.03st)
Condition:. . . . . .Cranial cruciate ligament
 rupture of the right knee,
 osteoarthritis of the right
 knee
Owned by:. . . . . .Jon Halls and Karen
 Ashton
Surgical procedures:. . .Tibeal plateau levelling
 osteotomy
Medication:. . . . . .Carprofen, a non-
 steroidal, anti-
 inflammatory and
 analgesic drug for short-
 or long-term pain.
 Tramadol, a non-steroidal
 and anti-inflammatory
 agent for mild to severe
 pain

Jon with Domino.

INTRODUCTION TO HYDROTHERAPY

Domino was showing stiffness and subsequent lameness following rest after exercise approximately four months prior to being examined by a vet. He was referred to a specialist veterinary hospital on 3 February 2009 for further investigation, and results from numerous radiographs and arthrotomy (surgical incision to the joint) revealed findings of a cranial cruciate rupture in the right knee. In addition, inflamed joints suggested secondary arthritic changes in the right knee. On 4 February, a tibial plateau levelling osteotomy procedure was carried out on Domino, effectively reducing the angle of the tibial plateau from 20 degrees to 9 degrees, with the aim of decreasing the angle to below 10 degrees to make it as stable as possible, and help prevent futher problems. Domino was discharged the following day.

However, despite successful surgery, complications can occur due to the trauma caused and recovery needed. Possible problems include implant failure and infection, as well as complications with surrounding bony structures like patella luxation, and fractures of the fibula and/or tibia. Jon and Karen were advised to rest Domino for six weeks, and only walk him in the garden on a lead for the first two weeks, whilst restricting him to a small room the rest of the time. Depending on his rate of recovery, he was then allowed five minute lead walks twice daily, increasing to three times a day if recovery allowed. At all times,

Domino's gait had to be assessed and any lameness immediately reported to the vet. Fortunately, Domino was recovering tremendously well and, on 16 March, Jon and Karen took Domino to the vet, for his six-week assessment.

Domino had been suffering from minor bouts of lameness due to his intense keeness to play ball, thereby over-exerting himself, but this was expected as he was unable to release pent-up energy through significant exercise. Slight lameness on the operated leg is normal at such an early stage of the healing process, and palpation of the limb was not painful. Radiographs revealed no sign of infection, the implants were in place, and the bone, though not fully healed, was developing well.

The vet suggested hydrotherapy in conjunction with strict controlled exercise at home to gradually increase Domino's strength, muscle mass surrounding

Domino has an increased chance of rupturing his left cruciate ligament as well as developing further complications.

the leg, and speed the healing process. Being a typical Spaniel, Domino loved any opportunity to swim, so Jon and Karen knew he would enjoy hydrotherapy and were enthusiastic to give it a try.

AIM OF PROGRAMME
Post-operative rehabilitation, maintenance of arthritis.

PROGRAMME DURATION
Domino began with two hydrotherapy sessions a week commencing March 2009 at no longer than five minutes a time. His sessions were reduced to one a week in January 2010, each 20 minutes in length.

TASKS TO BE COMPLETED AT THE CENTRE
Attend once a week for a 20 minute session wearing non-buoyancy jacket.

TASKS TO BE COMPLETED AT HOME
After his initial follow-up consultation with the vet, Domino's exercise was increased to ten minute walks on the lead, three times a day for a week. If there was no sign of lameness after this, he could walk for 15 minutes, three times a day each week until a further check-up with the vet. At present, Domino is exercised for 25 minutes twice a day off the lead, with an additional weekly 60-minute walk. As Domino swims once a week, on the day of his session, he has just one walk to avoid over-exercising.

Domino is not allowed to jump up or down, or run up and down stairs. His weight has never been an issue, and has remained stable throughout his life, meaning that related health problems have been and are minimised, to an extent.

PROGRESS REPORT
Hydrotherapy commenced in March with Domino attending two sessions a week in the pool in a buoyancy jacket, for no longer than five minutes at a time with a tennis ball (Domino is a toy-orientated dog, so it was decided that he would be more comfortable and relaxed with a ball). Domino made progress and continued to swim until May.

It was suspected that Domino's left knee would suffer the same cruciate trauma, and his programme was put on hold until further notice. On 3 November,

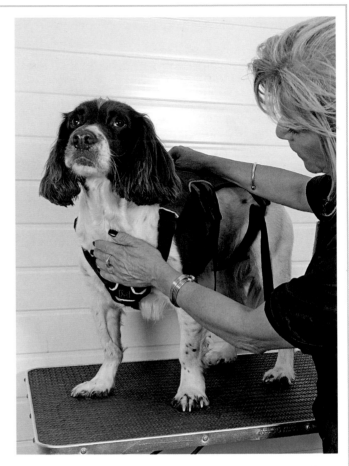

Domino wears a non-buoyancy jacket to increase the intensity of his programme.

Domino returned to continue his sessions twice a week, building to 8-10 minute sessions.

On 16 January 2010, Domino had built his stamina and strength to a high level. Hydrotherapists and vets were equally pleased with his improvement, and his sessions were reduced to once a week for fitness and maintenance. Domino's walks were still controlled but he was no longer restricted in the house. In February, the jets were added to Domino's programme to make the sessions more challenging, and continued for a month.

In May, Domino's programme was modified again. As his land-based exercise increased, he needed less of a challenge in the pool, so hydrotherapy was used as a maintenance form of exercise as Domino was allowed more vigorous walking and running activities. Domino was swimming without the jets for sessions of up to 20 minutes. Without the jets, Domino was able to swim much more steadily, but the challenge was no longer there. In May (up until present-day), Domino began wearing a non-buoyancy jacket which encourages him to work hard, adding a degree of challenge without over-exercising him.

Domino has not suffered any further problems with his right knee, and is in good health. Jon and Karen have noticed that Domino's muscle tone has become more defined, and his gait relatively even, with equal muscle mass on both hindlegs. They are pleased with the rehabilitation of his operated leg, and believe that hydrotherapy is maintaining/slowing the process of any complications with the left leg, as well as relieving pain caused by arthritis in his right leg. For this reason, Domino will continue hydrotherapy for the rest of his life.

Domino is a typical Spaniel; he'll do anything for a tennis ball!

Visit Hubble and Hattie on the web: www.hubbleandhattie.com and www.hubblendhattie.blogspot.com
Info about all H&H books • New book news • Special offers

93

11
Spinal conditions

The nervous system can be divided into two main parts: central and peripheral. The central nervous system consists of the brain and spinal cord, and the peripheral nervous system contains many nerve types that carry information from body structures to the sensory nerves. These are responsible for controlling most of the functions in the body, including balance, posture, gait, and muscular contraction.

The spine is made up of five sections: the cervical (neck), thoracic (mid back), lumbar (lower back), sacral (pelvic region), and coxygeal vertebrae (tail). Within each vertebra lie intervertebral discs that connect the bones of the spinal column, provide flexibility between the vertebrae, and act as shock absorbers along the spinal column. They have a soft centre (nucleus pulposus) inside a ring (annulus fibrosus). As a dog matures, the ring deteriorates, leading to reduced movement and shock absorption.

An intervertebral disc may slip in one of two ways. The soft centre can emerge from the ring and injure the spinal cord, which is known as extrusion. Conversely, the ring may thicken and compress the spinal cord to cause what is called a protrusion.

Sensory nerves in the spinal cord can become compressed and inflamed due to trauma, injury or disease. The animal will react to this by showing clinical signs in the shape of pain responses.

Many conditions in the dog can affect the spinal cord, some of which are neurological and others orthopaedic. These involve injury (eg fracture), disease, and neoplasia (cancer).

Spinal conditions occur either chronically or acutely from trauma, degeneration, genetic predisposition, or an unknown cause. All breeds of dog can be affected; however, certain breeds, like the Dachshund, Doberman and German Shepherd, are genetically predisposed to spinal disease.

Prognosis can be difficult as clinical signs can

The spinal cord is a complex and delicate structure. (Courtesy Dave Russell)

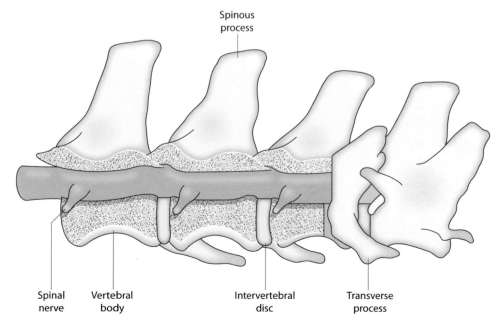

Spinous process

Spinal nerve

Vertebral body

Intervertebral disc

Transverse process

age, and can also be affected by injury or surgery, especially in relation to neurological diseases.

Walking on different surfaces stimulates sensory and proprioceptive nerves, which improves all functions on land. Altering the surface walked on (gravel, sand, uneven surfaces in water of various levels) encourages the body to adapt accordingly. Changes in direction such as figures of 8, around obstacles, and when walking backwards requires balance, control and spatial awareness.

Changing direction in the hydrotherapy pool offers resistance for the animal to work against, and requires more strength and endurance, which is a great way to increase difficulty once recovery has reached a certain level.

A dog's body weight is important in relation to recovery, especially so with spinal conditions as excess weight and lack of muscular fitness can lead to disk degeneration. Normal weight-bearing in a healthy dog puts enough strain on the spine without adding excess weight to the equation. Chondrodystrophic breeds (those with a long spine and short legs) can suffer further damage with the onset of osteoarthritis, tissue injury (ligament tear, sprain), and prolonged recovery.

In this chapter, the following conditions are discussed –

- intervertebral disc disease
- degenerative myelopathy
- fibrocartilaginous embolism
- discospondylitis/spondylitis
- cervical vertebral instability/cervical spondylopathy

INVERTEBRAL DISC DISEASE
Originates in the neck or mid to lower region of the spine, and is often referred to as a 'slipped disc.'

Two types of disc degeneration fall under this heading – 'Hansen I' and 'Hansen II,' which can occur individually or simultaneously. chondrodystrophic breeds, usually occurs suddenly (peracutely), or within a short period of time (acutely), and involves disc extrusion. It can occur in multiple discs of the cervical or thoracolumbar

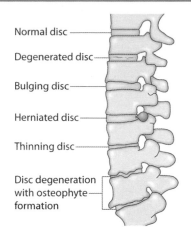

Normal disc
Degenerated disc
Bulging disc
Herniated disc
Thinning disc
Disc degeneration with osteophyte formation

Palpation, radiographs, myelograms, and CT and MRI scans all help eliminate certain spinal conditions, and get to the source of the problem. (Courtesy Dave Russell)

region. Hansen II is a slow (chronic) and ominous disc degeneration most commonly seen in non-chondrodystrophic breeds, resulting from protrusion.

Any dog can develop invertebral disc disease of the neck or mid to lower spinal region, and the clinical signs shown will depend on the amount of disc material involved, and the force with which the herniation occurs. When a cervical or thoracolumbar disc herniates, it causes damage to the nervous system, which initially causes pain. So, for example, paraspinal hyperesthesia (increased sensitivity to touch alongside the spinal column) is the most reliable clinical sign of disk herniation. If the neck region is affected, a dog will carry his head low and be reluctant to lift it, and have difficulty eating and lying down.

Invertebral disc disease can be treated surgically and non-surgically, according to severity. Mild pain and less severe clinical signs can be managed with strict rest, anti-inflammatory medication, and limited/no exercise for a period of time to watch how matters progress.

Decompressive surgery is the most common procedure, whereby a section of bone is removed from around the spine (laminectomy) to enable disc material to be removed. In a hemilaminectomy, a segment of disc material from the spine is removed from one side. Further slipped discs can be prevented by carrying out a disc fenestration, a surgical technique in which the soft centre of the invertebral disc is removed through a surgically created window (usually performed at the time of decompressive surgery).

Recovery from invertebral disc disease, though a relatively lengthy affair, is, overall, successful. To limit its incidence, it's important to ensure your dog does not suffer body trauma and avoids extreme exercise/activities.

DEGENERATIVE MYELOPATHY

The term 'myelopathy' is used to describe any functional disturbance or change in the spinal cord, and degenerative myelopathy is a slow, neurodegenerative disease. A dog with degenerative myelopathy will experience chronic deteriorating pelvic limb weakness, causing knuckling or dragging of the back feet, and a lack of co-ordination that will make it difficult to climb stairs. Degenerative myelopathy has similar symptoms to severe osteoarthritis, and some other spinal diseases. Unfortunately, degenerative myelopathy symptoms gradually worsen, either steadily or in phases, until, eventually, the dog's rear end is completely paralysed.

Provisional diagnosis is based on symptoms, history, breed, and age of the dog. Degenerative myelopathy can occur with intervertebral disc disease and inherited diseases such as hereditary myelopathy (in Afghan Hounds), but its actual cause is unknown. It commonly affects German Shepherds, Rhodesian Ridgebacks, and the Pembroke Corgi, as well as other breeds typically when over 8 years old. As the origin is not known, breed susceptibility and occurrence cannot be assumed.

Diagnosing degenerative myelopathy is by a process of elimination. DNA tests are now available to try and detect a mutated gene, but the most successful prognosis is, unfortunately, post-mortem.

No treatment exists for degenerative myelopathy, but it is obviously important to provide comfort and improve the wellbeing of the dog. Dogs diagnosed with degenerative myelopathy are usually euthanised at between six months and a year after developing the disease, as deterioration is such that their quality of life is affected, even though the disease is not painful. Typically, a dog with degenerative myelopathy will lose awareness and functionality of his physical capabilities.

To maintain muscle, help with stability, and attempt to slow the process, physiotherapy and hydrotherapy are prescribed.

FIBROCARTILAGINOUS EMBOLISM

Fibrocartilaginous embolism simply means a blockage (embolism) of fibrous tissue and cartilage (fibrocartilaginous) in the spinal blood vessel. The blockage is thought to be a result of intervertebral disc herniation, which blocks the blood vessel, causing restricted blood supply (or ischemia) to the spinal cord. A blockage can be present in either the arterial or venous blood supply of the spinal cord, or sometimes both.

Blood supply to the spinal column is by way of a complex structure of arteries and veins, and nourishes the cells in the vertebrae, spinal cord, nerves, muscles, and other structures. Disruption of the blood supply can result in infarction; an area of tissue which has died due to a lack of oxygen. Fibrocartilaginous embolism can be present in any part of the spine, although it occurs particularly frequently in the neck and mid to lower back regions.

Fibrocartilaginous embolism often occurs in association with vigorous exercise or mild trauma, typically affecting non-chondrodystrophic breeds. It has identical histopathology (the study of diseased tissues at a microscopic level) to the soft centre of intervertebral disc disease, but the cause is not known. It is a non-progressive condition, the symptoms of which are non-painful mental and/or physical function deficiencies, and often leads to ischemic myelopathy, a blockage to the spine.

No specific measures can be taken to prevent fibrocartilaginous embolism occurring as its cause is not fully understood. Physiotherapy and hydrotherapy are carried out as supportive care to train and improve spatial awareness and rebuild strength and mobility.

Fibrocartilaginous embolism often resolves itself over several weeks although some dogs do not fully recover, being left with an inability to feel pain in their feet, or a reduction in spatial awareness.

SPONDYLITIS

Spondylitis or discospondylitis is an infection of the spinal vertebrae and intervertebral discs, a type of vertebral bone infection that is sometimes referred to as a form of arthritis, as the condition destroys and narrows the discs of the spine.

Spondylitis is caused by septicaemia, an infection of the blood which can derive from bacteria that has entered through a wound, bladder infection, or immune disorder. A dog that is recovering after spinal surgery, for example, may have spinal trauma, which can bring about spondylitis as the body's defence system is reduced and is less able to fight off the bacteria. In some cases, the body does manage to get rid of the bacteria itself, although this is rare.

The bacteria travel in the blood stream and circulate round the body, making their way to the soft tissue areas (in the disc space) of the vertebra, where they multiply, resulting in dead tissue and bone destruction. Consequently, the spinal vertebrae become inflamed and sensitive.

All breeds of dog can be affected by spondylitis, although it is commonly seen in large breeds and middle-aged dogs. A dog suffering from spondylitis will exhibit progressive stiffness, weakness, fever, and depression. Any part of the spinal column can be infected, although it's the mid to lower back region that is the most frequently affected area. As the infection develops, the symptoms will become more severe, and can result in paralysis and parapesis.

Spondylitis is a chronic and slow-progressing condition that has to be treated immediately. In conjunction with a neurological examination, radiographs are the best way to diagnose spondylitis as the dead tissue can be seen and analysed in the areas affected.

Treatment involves antibiotics, analgesics (possibly with a need for decompressive surgery), and procedures to stabilise the spine, where necessary. Hydrotherapy can help an affected dog regain his strength and assist in recovery. Full recovery is dependant on how quickly the condition is diagnosed, and its severity.

Spondylitis can also be caused by fungal agents that cause a chemical reaction or change within the body, and, though rare, this type of discospondylitis is incurable, although it can be managed for some time.

CERVICAL VERTEBRAL INSTABILITY

Cervical vertebral instability, or cervical spondylopathy, is an abnormality of the vertebrae in the neck (spondylopathy is used to describe any disease of the vertebrae). Cervical vertebral instability can be due to genetics, nutrition (too much calcium), trauma, or a combination of all. Large and giant breeds are most affected, specifically the Doberman and Great Dane. The condition can occur at any age, and symptoms include 'wobbling;' hence its familiar name 'Wobbler's syndrome.'

Cervical vertebral instability is caused by compression of the neck vertebra as a result of malformation. These abnormalities cause major stress to the intervertebral discs, which subsequently begin to degenerate and rupture, creating additional pressure on the already compressed spinal cord. The compression damages the neck area of the spinal cord, which is needed to allow an animal to stand and move normally.

Consequently, an affected dog may show stiffness, a reluctance to move his head, be unwilling to eat (as he is unable to extend his neck), and feel and exhibit pain when the area is touched.

As the area affected is small, symptoms are not evident in the early stages. As the disorder progresses, the gait becomes unsteady, and the dog may even fall over.

Cervical vertebral instability is diagnosed and its severity determined via myelogram and radiographs. Medication to reduce inflammation and pain is frequently administered in mild cases. Where the condition is more severe, surgery is usually required to stabilise the region.

Aquatic exercise can be used post-surgery to aid recovery, but is rarely ever used as a form of conservative management. When swimming, a dog's head obviously has to be above the surface, so there is a high risk that hydrotherapy will cause more damage. Also, as some dogs react to entering the water by panicking, thrashing and making sudden movements (ie jolting of the head), this also adds to the danger. No real evidence exists to show that hydrotherapy is beneficial for this particular condition, so is not advisable.

Case history

Name:. . . . . .Duchess
Breed:. . . . . .Japanese Akita
Age: 7
Sex: . . . . .Female
Weight (at start):. 36.7kg (5.77st)
Condition: . . . . .Invertebral disc disease
Owned by:. . . . . .Nicole Saveall
Surgical procedures:. . .Hemilaminectomy
Medication:. . . . . .Gabapentin and Rimadyl. Gabapentin is an inhibitory neurotransmitter used to treat neuropathic pain. A neurotransmitter is a chemical released by neurons which allow information (impulses) to be passed from one cell to the next. Gabapentin works by delaying the impulses (pain of neurological origin), providing pain relief in the process

INTRODUCTION TO HYDROTHERAPY

The Saveall family rehomed two-year-old Duchess in 2005. She had no previous medical problems and led an active life. On 19 August 2010 she required an emergency neurological investigation as she was in severe pain. The examination revealed acute onset of non-ambulatory paraparesis, which was worse on the left side.

Duchess was showing a typical Schiff-Sherrington posture, deficient postural reactions, and heightened spinal reflexes in the pelvic region. An MRI scan of the spinal cord was carried out which revealed intervertebral disc extrusion of the lumbar 2 to lumbar 3 vertebrae, with spinal cord compression (Hansen I).

An immediate hemilaminectomy with fenestration (removal of a significant amount of disc material) was

continued over

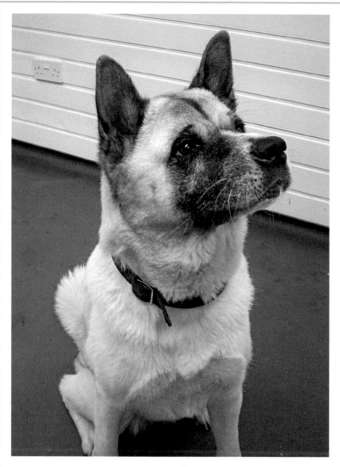

Japanese Akitas are not usually prone to intervertebral disc disease.

carried out. Duchess recovered well, and was soon able to urinate without assistance so was discharged (the most important consideration in post-operative care after a hemilaminectomy is to ensure that the bladder is regularly emptied, as urinary retention is the most common problem in disc disease of the mid to lower back).

The referral centre suggested that, ideally, Duchess should undergo hydrotherapy. Duchess' owners were willing to try anything that could help speed recovery, and get her back to her old self. The vet suggested that hydrotherapy should begin four to six weeks after her discharge, and once her stitches had been removed and her wound had healed.

Duchess responded well to treatment, and surprised her owners by tolerating the exercise. Nicole was pleased by the improvement in her posture and gait over such a short period of time, and wanted to continue treatment in the hope of further improvement in mobility, muscle development, stamina, and strength.

AIM OF PROGRAMME
Post-operative rehabilitation to help develop and maintain muscle strength and spinal stability.

PROGRAMME DURATION
Duchess began water therapy on 30 September 2010, starting with three minutes in the hydrotherapy pool, three times a week. It was intended that duration and intensity would gradually increase alongside her physiotherapy.

TASKS TO BE COMPLETED AT THE CENTRE
Swim in a buoyancy jacket with a hydrotherapist

Patients recovering from spinal surgery often display a hunched posture.

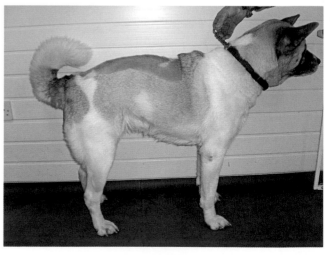

When she first began hydrotherapy, Duchess required support of the pelvic region and encouragement of her hind limbs.

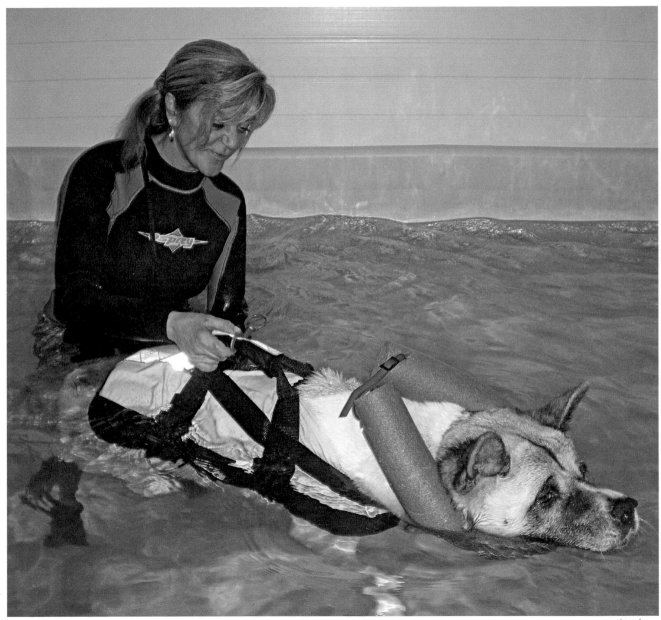

continued over

supporting her pelvis, massaging and encouraging her hindlegs, and ensuring she swims in a straight line to prevent twisting of the spine and over-rotation of her hips. To keep Duchess' head from dropping into the water and also keep her spine aligned, a head flotation ring was used to support her.

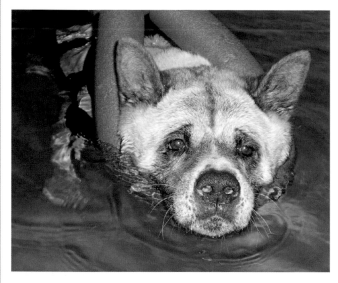

To keep her spinal column in line, a floatation device is used to lift her head slightly.

TASKS TO BE COMPLETED AT HOME

The Saveall family was advised to not allow Duchess to run, jump or go up and down stairs for at least three months, and ideally for the rest of her life. Non-slip surfaces throughout the house were advised to prevent further injury. Duchess is allowed short and frequent, 10 to 15 minute walks on a harness and lead three times a day. To develop her awareness, Duchess is walked slowly over various surfaces (grass/gravel, etc) as it will make her use and fine tune muscle control to adapt to different terrains.

Physiotherapy was carried out two to three times a day whilst Duchess was hospitalised. Strength, endurance, and balance exercises were to be done at home; for example, walking Duchess in small and large circles, clockwise and anti-clockwise, figures of 8, walking up gentle slopes to improve strength and balance, and transition from sit, down, and stand, whilst offering support when required.

PROGRESS REPORT

Duchess is more tolerant and accepting of being handled by different people, from an initial wariness of strangers. Duchess has become much more stable and exhibits a more relaxed and even, and less stiff gait. As a result, the Schiff-Sherrington posture is less evident and she is not as 'hunched.'

In September 2010 Duchess underwent a neurological reassessment. Results showed a marked improvement in her condition, and Animal Magic was advised to steadily increase her exercise programme over the next few months before re-examination in November, or prior to that if she showed signs of deterioration. To make her programme slightly more challenging, Duchess' swim duration was increased to 10 minutes, three times a week.

After her second reassessment, Duchess was still recovering exceptionally well and coping with the increase in exercise. Subsequently, both physiotherapy and hydrotherapy sessions were increased in difficulty

New strength in her paraspinal and quadriceps muscles means she now swims unassisted, without encouragement to move her legs.

with circling and figures of 8 more frequent and intense (small circles and figures of 8 require tighter bends and turns).

Nicole has successfully maintained Duchess' weight, and improvement in her fitness is reflected at home in increased energy levels and more adventurous characteristics. To increase the intensity of her hydrotherapy, on 4 January 2011, Duchess swam unassisted for the first time, and continues to do so.

Complementary therapies have achieved remarkable improvement in Duchess' condition, and it is likely that aquatic exercise will be needed to maintain her fitness as land-based exercise puts unwanted pressure on the spine, even though her paraspinal muscles have developed for support.

Case history

Name:. . . . . Bertie
Breed:. . . . . German Shepherd
Age:. 8
Sex:. . . . . .Male
Weight (at start):. 32kg (5.03st)
Condition:. . . . . .Ischemic myelopathy
Owned by:. . . . . .Andy Johnson
Surgical procedures:. . .Triple pelvic osteotomy
Medication:. . . . .Carprofen, Gabapentin

INTRODUCTION TO HYDROTHERAPY

Andy acquired Bertie, his loyal companion, when Bertie was 10 weeks of age. When Bertie was 11 months old, a successful triple pelvic osteotomy was performed to correct his right hip, as he had mild dysplasia. The surgery also helped prevent this from progressing to severe dysplasia, which is common in German Shepherds. At the time, hydrotherapy was not suggested as an aftercare treatment, although, once recovered, Bertie did swim in leisure pools designed for dogs and owners (in Japan) once a month.

Bertie led a very active life, so it came as a big shock when, one morning in March 2010 (in England), Bertie could not get up from his bed, let alone walk. After being seen immediately by a vet, an MRI scan revealed that Bertie had a fibrocartilagenous embolism of the thoracic vertebrae 3 to lumbar vertebrae 3, which is what caused the acute ischemic myelopathy (lack of blood supply to the spine) and Bertie's paralysis.

Bertie was admitted to a veterinary hospital for

Andy with Bertie.

continued over

Bertie is a football enthusiast!

two weeks, and underwent extensive physiotherapy. On discharge from hospital, Bertie was able to walk, although his gait was very unstable. Bertie was on medication for a total of one month; 2 weeks whilst in hospital and 2 weeks after discharge.

He received attention from a physiotherapist, and at home with Andy. Bertie had restricted movement in his hindlegs and, not surprisingly, was unable to lift his feet off the ground to walk over a step and replace them in the right position. As Bertie had limited mivement in his hind legs, the physiotherapist used passive exercise (whereby he actually moved Bertie's limbs), as well as mobility, balance and strengthening exercises. One of the exercises was proprioception training, which entailed Bertie walking on and around a track that was covered with various surfaces and obstacles in order to enhance his body awareness and movement control. In addition to his rehabilitation programme, his vet recommended that Bertie should start hydrotherapy sessions.

AIM OF PROGRAMME
To build muscle around the hips and hindlegs to support his weight and improve stability, as well as improve mobility and strength. Hydrotherapy was provide non-weight-bearing exercise in addition to physiotherapy and controlled, land-based exercise to speed the recovery process and promote a sense of wellbeing.

PROGRAMME DURATION
Hydrotherapy sessions began on 3 April 2010 and continued for 8 months of thrice-weekly sessions.

TASKS TO BE COMPLETED AT THE CENTRE
Swim for between 12 and 15 minutes, dependent on other exercise that day. If Andy takes Bertie out for a long walk, for example, then the session that day is reduced accordingly.

Bertie loves all sorts of toys, from chasing remote control cars to playing football. This knowledge is used to ensure he is comfortable and enjoys hydrotherapy by allowing him to hold a toy throughout the session; he becomes distressed without it. Bertie swims in a buoyancy jacket attached to two lead ropes, with the hydrotherapists outside the pool. The lead ropes are attached to the rear 'D' ring to provide support for his lumbar region.

TASKS TO BE COMPLETED AT HOME
Twice daily Andy performs Tellington Touch massage

Walking on sand and over varying terrain helps improve proprioceptive function (spatial awareness).

Aquatic exercise has improved Bertie's overall body condition, and strengthened his thoracolumbar region.

on Bertie, which involves massaging in circles using the thumb and forefinger. To help Bertie with mobility, a cavaletti (a small, portable jump) is used to refine his co-ordination. This exercise particularly helps to ensure that dogs affected by neurological problems concentrate on their movement, focusing on where they are placing their hindlegs, thus reducing gait abnormalities and promoting balance.

PROGRESS REPORT

Bertie can now swim for 15 minute sessions. He started off swimming 7 minutes at a time; his previous experience in swimming pools, and his overall health and fitness helping him increase his swim duration quickly.

When Bertie first came to Animal Magic he was physically weak and very unstable, his stance and walking gait 'wobbly,' due to muscle atrophy and trauma of the spinal column. Bertie has improved dramatically, and is now able to run up and down stairs and play football, although Andy is very careful with how much exercise Bertie does.

Bertie's muscle development has improved greatly; reflected not only in his activity levels but also his weight and increased muscle mass. Bertie's weight at its heaviest was 38.40kg (6.04st), but is now maintained

at around 37kg (5.82st) to keep strain on his body to a minimum. Excess weight plus spinal conditions plus previous hip problems equals additional pressure and stress.

Andy has noticed that Bertie's energy levels and physical ability are greatly improved, and is pleased with his rehabilitation so far. Bertie will continue hydrotherapy to maintain his health, as well as slow the onset of arthritis.

Rope attachment to the pelvic area provides Bertie with extra support.

Case history

Name:. . . . .Chester
Breed: . . . .Lhasa Apso
Age:. 5
Sex:. . . . .Male
Weight (at start): 10kg (1.57st)
Condition: . . . .Intervertebral disc disease
Owned by:. . . . .Heather Ross
Surgical procedures: . ..Hemilaminectomy
Medication: . . . .Methocarbamol; a muscle relaxant used to treat muscle spasms arising from disc disease

Heather with Chester.

INTRODUCTION TO HYDROTHERAPY

On the morning of 2 June 2010, Chester's hind limbs appeared paralysed, and he was unable to get up from his bed. Heather rushed him to a veterinary specialist for examination where a MRI scan was carried out. The results of this confirmed a prolapsed intervertebral disc; thoracolumbar intervertebral disc disease.

Chester underwent an immediate hemilaminectomy (removal of a section of bone over the spinal cord) to relieve pressure there, and hospitalised for ten days, and given muscle relaxants and pain relief.

To help with his rehabilitation, hydrotherapy was advised. Heather was unaware of hydrotherapy as a canine treatment, but felt confident that if her vet recommended it, it must be worth a try.

AIM OF PROGRAMME

To improve paraspinal muscles, fitness, overall body condition, and strength, and as a way of allowing Chester to exhaust his high energy level.

PROGRAMME DURATION

Chester began water therapy on 16 June 2010, attending twice a week.

TASKS TO BE COMPLETED AT THE CENTRE

To swim in a buoyancy jacket with ropes attached to the rear 'D' ring to allow support of the pelvic region.

TASKS TO BE COMPLETED AT HOME

No running or jumping for at least three months

When he first attended hydrotherapy, Chester was swum over the side of the pool.

Chester developed conformational abnormality in his front legs as he grew.

following surgery, and ideally no jumping for the rest of his life to avoid additional stress on the spinal column.

PROGRESS REPORT

When Chester was first introduced to the pool, he was nervous and panicky, but gradually learned that the water was a safe environment, and became relaxed enough to play with a tennis ball occasionally, which gave him confidence.

Chester began his post-operative programme with 4-minute water exercise sessions. He coped with hydrotherapy really well and, from July 6, 2010, was physically fit enough to attend just once a week. By 14 October, he was swimming for 14 minutes a session, with rest periods. As a result of Chester's hydrotherapy sessions, he became more toned and appeared to have lost weight. His weight was maintained at around 9.5kg (1.49st), which is more of an ideal weight for a dog of his size and condition in order to prevent extra strain to his body.

Whilst all his goals were being achieved, Chester's high energy levels were now the main concern as restricted exercise at home was causing him much frustration; increased energy levels meant that

continued over

As his confidence grew, Chester swam on rope attachments.

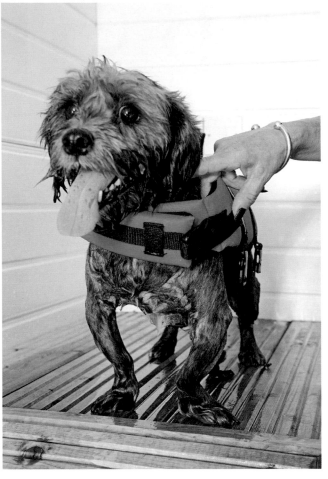

Due to the speed at which he swam, Chester required several rest periods to enable him to recover.

Chester wanted to jump and run around the house. Hydrotherapy helped in the early part of Chester's recovery, and, as he was able to carry out more land-based exercise, on 23 September 2010, his sessions were stopped.

12
The future

There are quite literally hundreds of disorders, diseases and injuries that can benefit from the effects of hydrotherapy. This book has covered only a selected few in order to advise and explain the positive outcomes that this ancient exercise is able to achieve. An overview of the most commonly diagnosed conditions, in conjunction with actual case studies, have hopefully given an insight into how and why this particular therapy is so successful, and increasing in popularity. Not only does hydrotherapy assist in the healing of injuries, it can also slow development of debilitating disorders such as osteoarthritis. As a natural and enjoyable form of exercise, water therapy is said to provide a workout four times more effective than land-based exercise. Thus, it promotes rapid improvement in cardiovascular fitness, boosts healing, burns energy and builds strength without the need to weight-bear and possibly strain joints.

Of course, not every condition can be helped with hydrotherapy, and, in some cases, it can have the opposite effect. Similarly, not all dogs are suitable candidates for this form of exercise as it can compromise their health. Veterinary advice and discussion with approved hydrotherapists can help ensure that the correct rehabilitative care is provided for your dog.

Animal Magic continues to administer hydrotherapy where possible, and advise about its benefits. An underwater treadmill is planned for the centre which, together with the pool, will offer partial to full weight-bearing and non-weight-bearing facilities, in conjunction with physiotherapy learning and training.

Successful hydrotherapy is achieved by a mixture of expertise, experience, patience and devotion on behalf of the hydrotherapists, who are specifically trained to provide the best and safest treatment possible to achieve the most successful outcome. Every dog is different, and so each patient requires a tailor-made aquatic exercise programme specifically designed to meet his or her needs and objectives.

We need to compensate for the mistakes made through poor and (at times), irresponsible breeding programmes; there are hundreds of canine disorders that have come about as a result of selective breeding and human intervention. Dogs are, unfortunately, unable to speak for themselves, but by observing the simple guidelines set out in the Five Freedoms, we can ensure our dogs enjoy better welfare. Fitter and leaner dogs live longer – it's a fact. Although many health conditions cannot be prevented (when genetically based, say), we can delay their onset and progression by providing our dogs with adequate nutrients and exercise.

It is common knowledge how devastating obesity can be, but we do know how to control weight in a sensible way, and prevent problems occurring. Overfeeding a dog to the point of obesity is not an act of kindness; it can cause pain and suffering, and result in a shortened life span.

In very many ways, when it comes to healing and wellbeing, water really can be magical!

For the majority of dogs, hydrotherapy leads to a brighter future For others, it helps them to live life to the full.

13
Appendices

Further reading & websites

Books

Canine Hydrotherapy: the use of water to help dogs with sore backs, hips, and joints, like the Golden Retriever, Boxer, German Shepherd, Greyhound Labrador, and more
by Dana Rasmussen Lightning Source UK Ltd (2011)
ISBN 9781241312336

Canine Rehabilitation and Physical Therapy
by Darryl Mills MS DVM, David Levine PhD PT, and Robert A Taylor DVM MS Saunders (2004)
ISBN 9780721695556

My dog has cruciate ligament injury – but lives life to the full!
by Kirsten Hausler and Barbara Friedrich Hubble & Hattie (2011) ISBN 9781845843835

My dog has hip dysplasia – but lives life to the full!
by by Kirsten Hausler and Barbara Friedrich Hubble & Hattie (2011) ISBN 9781845843823

My dog has arthritis – but lives life to the full!
by Gill Carrick Hubble & Hattie (2012)
ISBN 9781845844189

The Complete Dog Massage Manual – Gentle Dog Care
by Julia Robertson Hubble & Hattie (2011)
ISBN 9781845843229

Exercising your Puppy: a gentle and natural approach – Gentle Dog Care
by Julia Robertson & Elizabeth Pope
Hubble & Hattie 2011 ISBN 9781845843571

Living with an Older Dog – Gentle Dog Care
by David A;derton & Derek Hall
Hubble & Hattie (2010) ISBN 9781845843459

Websites

Canine Hydrotherapy Association
www.canine-hydrotherapy.org/
A UK-based organisation whose aim is to provide
self-regulation and set benchmark standards
of treatment, operation, training, supervision,
first aid, record keeping, and water quality for
all our members. Also conducts research into
canine hydrotherapy and welcomes enquiries
or contributions from any interested party in this
respect

In this way veterinary surgeons, pet insurance
companies and – most importantly the dog owner
– can use a CHA member pool with confidence.

Canine Hydrotherapy Australia
www.caninehydrotherapy.com.au/

National Association of Registered Canine
Hydrotherapists (NARCH)
www.narch.org.uk/
The National Association of Registered Canine

Hydrotherapists (NARCH) is a not-for-profit
professional association which maintains the List of
Registered Canine Hydrotherapists (RCHs) in the UK.

NARCH ensures that all RCHs adhere to the
highest professional standards and ethics and
these requirements are laid out on this website
and within the Guide to Professional Conduct for
Registered Canine Hydrotherapists.

Association of Canine Water Therapy
www.iaamb.org/
The Association of Canine Water Therapy is a
division of the International Association of Animal
Massage & Bodywork, and is dedicated to
advancing the safe practices of canine water
therapy through education, establishing industry
standards, and building a network of support.

Visit Hubble and Hattie on the web: www.hubbleandhattie.com and www.hubblendhattie.blogspot.com
Info about all H&H books • New book news • Special offers

115

Glossary

Acute
Of or relating to a disease or a condition of rapid onset, with severe and recognisable symptoms, but often a quick recovery after treatment

Analgesic
A medication used to manage, reduce or eliminate mild to moderate pain, usually by acting on the central nervous system

Anterior
Situated at or directed toward the front or relating to the front surface of the body

Anthropomorphism
The attribution of human motivation, characteristics, or behaviour to non-human organisms or inanimate objects

Aquatic exercise
A physical activity that is planned, structured, and repetitive for the purpose of conditioning any part of the body. The subject (such as a dog) is totally or partially emerged in a water-based environment

Arthritis
Inflammation of a joint, usually accompanied by pain, swelling, and stiffness, resulting from infection, trauma, degenerative changes, metabolic disturbances, or other causes

Balneotherapy
The practice of administering medicinal baths in order to treat injury or disease

Bromine
A liquid element used in disinfecting water and in various pharmaceuticals

Capillary refill time (CRT)
Test used for quick assessment of shock or circulatory failure. Pressure applied to the gum causes blanching; normal CRT is around 2 seconds

Cardiovascular
Relating to the heart and blood vessels

Cartilage
Tough, elastic, connective tissue found in various parts of the body, such as the joints

Cartrophen
The brand name for an injection made from an anti-inflammatory drug (pentosan polysulphate) used in the treatment of canine and feline osteoarthritis

Caudal
Relating to or near the tail or hind parts of an animal

Central nervous system
The portion of the vertebrate nervous system that includes the brain and spinal cord

Chlorine
A common gas used to purify water, and as a bleaching agent and disinfectant

Chronic
Relating to an illness or medical condition that is characterised by long duration or frequent recurrence

Clinical condition
A medical condition of a part, organ, or system of an animal. It can result from various causes, such as infection, genetic defect, or environmental stress. It is characterised by an identifiable group of signs or symptoms such as ill health

Complementary therapy
A treatment or therapy used alongside and in support of conventional treatment

Congenital
Condition that is present at birth, as a result of either heredity or environmental influences

Contraindication
A condition which makes a particular treatment or procedure inadvisable

Cranial
Of or relating to the skull or cranium

Cruciate
Shaped or arranged like a cross; usually applied to the ligaments between the femur and tibia

Dam
Female parent of an animal

Diabetes mellitus
Metabolic disease where carbohydrate, fat and protein metabolism is defective owing to a failure to produce or respond to insulin

Electrotherapy
A treatment in which electric current is passed through the body tissues to stimulate muscle function. The current produces muscular contractions as an aid to muscular re-education following injury, or when suffering from certain neurological conditions

Epilepsy
A chronic nervous disorder characterised by sudden and complete loss of consciousness, associated with muscular convulsions, or fits

Ethics
Moral principles that govern a person's or group's behaviour

Extracapsular iliofemoral sutures
This surgical technique requires suture material to be placed in a figure of eight pattern from a part of the hip (ilium) to the thigh bone (femur)

Femoral shaft
The mid-section of the thigh bone (femur)

Fibre
A thin, threadlike structure

Fibrous
Consisting of, containing, or resembling fibres

Gait
A particular way or manner of moving

Geriatric
Of or relating to aging

Hereditary
Passed or capable of being passed from parent to offspring by genes

Hydrodynamics
The study of forces that act on or are produced by liquids

Hydrotherapy
The internal and external use of water in the treatment of medical conditions, disease and injury

Hyperadrenocorticism
Excessive production of cortisol (a hormone), more commonly known as Cushing's Disease

Hypoadrenocorticism
Partial or total failure of the function of the adrenal glands, which are located above each kidney. Also known as Addison's Disease

Iliac wings
Part of the hip bone (pelvis)

Immunosuppressant
A drug that reduces the body's normal immune response

Joint
The junction between two or more bones or elements of a skeleton

Labrum hip surgery
The labrum is a type of cartilage that surrounds the socket of the hip joint. It helps provide stability to the joint by deepening the socket and allowing flexibility and motion. It can become torn and may need surgery to smooth the edges

Lameness
Gait abnormality caused by pain or mechanical restrictions

Laminitis
Inflammation of the sensitive laminae in the hoof

Laryngeal paralysis
An acquired or congential condition in which the nerves and muscles that control the movements of cartilage of the larynx (a part of the respiratory tract) cease to function

Lateral
Toward the side, away from the midline

Leptospirosis
A zoonotic infectious disease caused by leptospira bacteria

Ligament
Tough, fibrous band connecting bones or cartilage at a joint, or supporting an organ

Luxation
Dislocationof a joint

Massage
The act of kneading, rubbing, etc, parts of the body to promote circulation, suppleness, or relaxation

Medial
Toward the midline; opposite to lateral

Metacam
Brand name for the anti-inflammatory drug meloxicam used to manage pain and inflammation

Mucous membrane
Protective layer of tissue that lines various canals and cavities of the body

Muscle
Contractile tissue that controls movement and locomotion; divided into three categories: smooth, cardiac, and voluntary

Muscle atrophy
A decrease in the mass of the muscle; partial or complete wasting away

Musculoskeletal
Organ system comprising the bones and their

attached muscles, responsible for the support and movement of animals

Neurological
Of or relating to the nervous system or neurology

Obesity
The condition of being obese; increased body weight caused by excessive accumulation of fat

Orthopaedic
The branch of medicine that deals with the prevention and correction of injuries or disorders of the skeletal system, and associated muscles, joints, and ligaments

Osteotomy
The surgical cutting or dividing of bone, usually to correct a deformity

Palpation
Feeling with the hand, applying light pressure to the surface of the body with the fingers to determine the condition of the organs beneath

Paraparesis
Partial paralysis of the lower limbs

Paraspinal
Adjacent to the spine (paraspinal muscles)

Parvovirus
A type of virus that causes contagious disease in humans, dogs and other animals

Passive bathing
Immersion in water whilst inactive

Passive exercise
Gently moving parts of the body (usually a joint) through their natural range of motion to improve flexibility and circulation

Peripheral nervous system
The part of the nervous system in vertebrate animals that lies outside the central nervous system. It includes the nerves that extend to the limbs and many sense organs

Physiology
The study of the functions of living organisms and their parts

Physiotherapy
The treatment of disease, bodily defects, or bodily weaknesses by physical remedies and modalities

Previcox
Brand name given to an anti-inflammatory drug which controls pain and inflammation caused by osteoarthritis in dogs

Prolapse
Displacement of a structure or organ from its usual location (prolapsed uterus)

Range of motion
The range, measured in degrees of a circle, through which a joint can be extended and flexed

Respiration
The act or process of inhaling and exhaling as the exchange of oxygen and carbon dioxide between the atmosphere and the body cells takes place

Rimadyl
Brand name given to an anti-inflammatory agent used to relieve pain

Ringworm
A common fungal infection characterised by circular lesions on the skin surface

Sarcoptic mange
A type of chronic skin disease of mammals caused by parasitic mites and characterised by skin lesions, intense itching, and loss of hair

Sire
Male parent of an animal

Subluxation
Incomplete or partial dislocation of a joint where there is still some contact between the bones' articular ends

Synoquin
Joint supplement used to help promote cartilage growth and maintain joint form and function

Tendon
Tough, fibrous band that connects muscle to bone or cartilage

Tendonitis
Inflammation of the tendon

Therapy
Treatment of illness or disability

Tramadol
Brand name for tramadol hydrochloride to relieve moderate to severe pain

Ultrasound
The use of ultrasonic waves to provide images, specifically for diagnostic or therapeutic purposes

Utilitarianism
The belief that the value of a thing or an action is determined by its usefulness

Von Willibrands disease
An inherited bleeding disorder caused by a deficiency or abnormality of a protein called von Willebrand factor, which is responsible for blood clotting

Welfare
Wellbeing of an animal

my
DOG
has
HIP DYSPLASIA
— but lives life to the full!

A practical guide for owners

Hubble & Hattie

ISBN 9781845843823

* Unique and practical guides with easy to understand text

* Step-by-step explanation of all phases of these conditions and their symptoms

* Understand what your vet advises

* Help your dog remain calm in a stressful situation

* Surgical options

* Aftercare options

* What to do next

* Motivate your dog to recover

* Discover how you and your faithful friend can enjoy life just as you always have!

* Case histories

**The *My Dog* ...
series**

**80 pages and
50-plus
illustrations,
all for just
£9.99* each!**

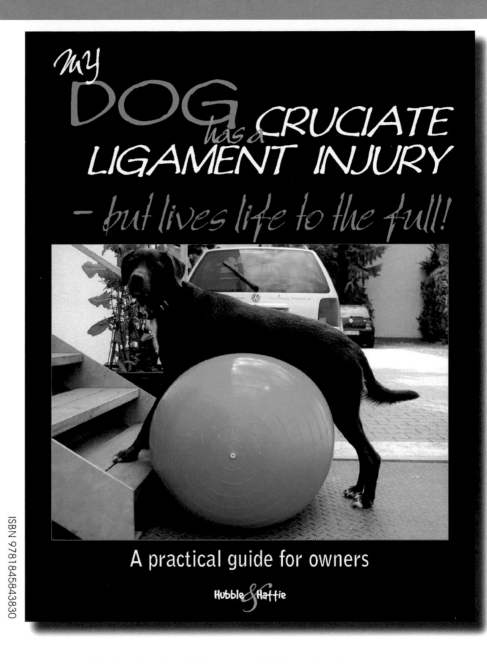

ISBN 9781845843830

my DOG has ARTHRITIS
– but lives life to the full!

A practical guide for owners

Hubble & Hattie

ISBN 9781845844189

Index

Acute ligament 85
Animal Magic 5-8
Anthropomorphism 15
Arthritis 40-42

Balneotherapy 18
Body condition score 51
British Veterinary Nursing Association 57
Buoyancy 35

Canine Hydrotherapy Association (CHA) 24
Capillary refill time 27
Cervical vertebral instability 100, 101
Complementary therapy 16, 17
Cruciate ligament 78, 85-93

Degenerative myelopathy 99
Density 35
Diet 53

Elbow dysplasia 71-76
Extracapsular 86, 87
Extracapsular iliofemoral sutures 65

Femoral head and neck excision 58, 59
Femoral head ostectomy 58, 59
Fibrocartilaginous embolism 99, 100
Five freedoms 14

Fragmented coronoid process 71, 73

Gait 28

Hip dysplasia 56-69
Horses 19
Hydrodynamics 35
Hydrostatic pressure 36

Intracapsular 86, 87
Invertebral disc disease 98

Kennel Club 57

Laryngeal paralysis 88

Mucous membranes 27

National Association for Registered Canine
    Hydrotherapists (NARCH) 24

Obesity 50-54
Osteoarthritis 40, 56
Osteochondritis dissecans 71, 73, 74
Osteochondrosis 71, 73, 74

Patella luxation 78-83
Pre-hydrotherapy assessment 26

Proprioception 96

Rheumatoid arthritis 41

Septic arthritis 42
Specific gravity 35
Spondiylitis 100
Synovial joint 40

Temperature, pulse and respiration 29
Tibial plateau levelling osteotomy 87, 90

Tibial tuberosity transposition 80
Total hip replacement 58-60,
Triple pelvic osteotomy 58, 74, 105
Trochleoplasty 80

Underwater treadmill 20-22
Ununited anconeal process 71, 72

Viscosity 36

Weight management 54

Visit Hubble and Hattie on the web: www.hubbleandhattie.com and www.hubblendhattie.blogspot.com
Info about all H&H books • New book news • Special offers

Notes